Lipids as a Source of Flavor

Lipids as a Source of Flavor

Michael K. Supran, EDITOR
Thomas J. Lipton, Inc.

A symposium sponsored by the Flavor Sub-Division of the Division of Agricultural and Food Chemistry at the 174th Meeting of the American Chemical Society, Chicago, Illinois, August 30–31, 1977.

ACS SYMPOSIUM SERIES **75**

AMERICAN CHEMICAL SOCIETY
WASHINGTON, D. C. 1978

Library of Congress CIP Data

Main entry under title:
Lipids as a source of flavor.
(ACS symposium series; 75 ISSN 0097-6156)

Includes bibliographies and index.

1. Lipids—Congresses. 2. Flavor—Congresses.
I. Supran, Michael K., 1939- . II. American Chemical Society. Division of Agricultural and Food Chemistry. Flavor Subdivision. III. Series: American Chemical Society. ACS symposium series; 75.

QC305.F2L46 664'.06 78-9739
ISBN 0-8412-0418-7 ASCMC8 75 1–121 1978

PRINTED IN THE UNITED STATES OF AMERICA

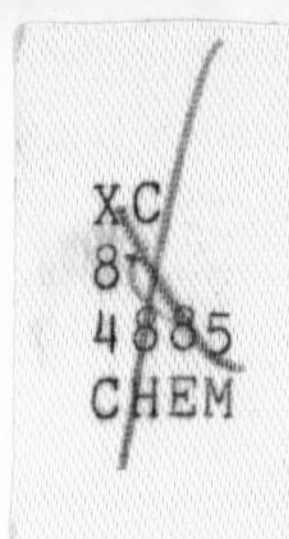

ACS Symposium Series

FOREWORD

The ACS SYMPOSIUM SERIES was founded in 1974 to provide a medium for publishing symposia quickly in book form. The format of the SERIES parallels that of the continuing ADVANCES IN CHEMISTRY SERIES except that in order to save time the papers are not typeset but are reproduced as they are submitted by the authors in camera-ready form. As a further means of saving time, the papers are not edited or reviewed except by the symposium chairman, who becomes editor of the book. Papers published in the ACS SYMPOSIUM SERIES are original contributions not published elsewhere in whole or major part and include reports of research as well as reviews since symposia may embrace both types of presentation.

CONTENTS

PREFACE

In organizing this symposium, I had the opportunity and the pleasure of enlisting some of the most eminent researchers in the field of lipid chemistry. Our objective was to review available information and to highlight current research in the field of "Lipids as a Source of Flavor."

As a product developer, I have a keen appreciation for the importance of our topic. In my experience, it is rare that a food product's quality, in a positive or negative sense, can be divorced from consideration of our subject.

I believe that attendees to this symposium found the program to be highly informative. I sincerely hope that the reader also will find it so, and therefore will join me in thanking both the authors for their research endeavors and the speakers for their expert presentations.

Thomas J. Lipton, Inc.
Englewood Cliffs, New Jersey
March, 1978

MICHAEL K. SUPRAN

1

The Role Lipids Play in the Positive and Negative Flavors of Foods

IRA LITMAN and SCHELLY NUMRYCH

Stepan Chemical Company, Flavor/Fragrance Division, 500 Academy Drive, Northbrook, IL 60062

As members of a group of flavor chemists, we have long had a special interest in those flavors that are derived from lipids, for lipids are generally associated with flavor defects in foods. It will not be a surprise to many of you to learn that lipids are also the source of many of nature's finest flavor creations.

Lipids, proteins, and carbohydrates, the chief structural components of living cells are also the major sources of flavor in foods. Of the three, lipids may be the most important for the following reasons:

1) They are precursors for many flavorful compounds, with representatives in the aliphatic aldehyde, ketone, lactone, fatty acid, alcohol and ester groups.

2) The intact glyceride tends to modify the flavor of fat soluble compounds by restraining their escape into the air space above.

3) They may also interfere with gustatory ingredients such as salts, sweetening agents, bitterants and acidulants from reaching saliva, a prerequesite for the sense of taste to occur.

4) As a cooking medium, triglycerides produce in foods special flavor effects as a result of, for example, deep fat frying. Their role here is to transfer heat and flavor to the cooking product.

5) It is frequently overlooked that lipids provide a range of polar and non-polar food grade solvents that are used by the food industry. From this range of solvents, one can select a suitable carrier for most any volatile material. Among these solvents are glycerol, mono- and diglycerides, triacetin, tributyrin and vegetable oils.

6) It is also important to note that glycerides, which act as reservoirs of volatile flavor, are themselves non-volatile and in this form they contribute to flavor largely through mouth stimulation. For example, they

0-8412-0418-7/78/47-075-001$05.00/0

impart to melting butter a characteristic "cooling" effect. When emulsified in milk, they impart a richness that is sorely missed in the flavor of skim milk. In ice cream, the solidified fat globules give the product its characteristic "creamy-dryness" associated with good quality.

Generally, the negative qualities in food flavor are associated more closely with lipids than with carbohydrates or proteins. Lipids are responsible for rancidity and oxidized flavors in beverage milk, butter (1) and vegetable oils (2) and for spoiling wet fish (3). They are involved in the stale flavors of potato flakes (4) and baked goods (1). Lipids are thought to be responsible for soybean reversion flavor (5), for warmed over meat flavor (6), for old heated cooking oil flavor (7), for the rancid flavors in peanuts, coconut, coffee and chocolate (8), and for many others.

On the other hand, lipids are also responsible for much of the desireable flavor of tangy cheeses such as cheddar and roquefort (9), for the flavor of fresh milk (1), for the "creamy" flavor of cream (1), for the "rich" flavor in heated butter (10) and for the characteristic flavors of mushrooms (11), green beans (12), peas (13), tomatoes (14) and cucumbers (15) and for much of the ripe flavor in fruits and berries.

The significance of lipids to odor may well begin at the site of olfaction, at the two and a half square centimeter patch of highly enervated tissue located at the roof of the nasal cleft. Here, volatile molecules are thought to be adsorbed and polar oriented between the lipid membrane portion of the nerve and the surrounding aqueous layer. It has been theorized that the adsorption and desorption of these molecules triggers an electric impulse which the brain interprets (16).

How important are lipids in artificial flavors? This question was answered in a recent survey of our flavor formulations. It was obvious that lipids are the most common ingredients used both in quantity and in variety. Only terpinoid compounds derived from the essential oils of spices, woods, citrus, and others compete in this regard. In the United States, there are presently 1150 volatile compounds permitted in artificial flavors. Of these, nearly one fourth (275) are, when found in nature, presumed to be derived from lipids. Forty five percent of these are esters, sixteen percent are aldehydes, thirteen percent are alcohols, nine percent are acids, nine percent are ketones and seven percent are lactones. Several artificial flavors are composed entirely of lipid compounds while most other compositions depend heavily on them.

In nature, how do these compounds arise? We know that oxidation of lipids is the first step. Autoxidation of polyunsaturated lipids is one of these mechanisms and is of concern to the oil chemist and the food technologists because it is a non-enzymatic and self-sustaining reaction that can cause off-flavor development, toxicity, and destruction of some oil soluble vitamins. During the storage of processed foods and oils,the formation of hydroperoxides and their decomposition products proceeds by way of free radical mechanisms. The production of free radicals is in turn promoted by external energy sources such as heat, light, high energy irradiation, metal ions, metallo-proteins such as heme, and others. These lipid hydroperoxides, the initial products of autoxidation, will, if left unchecked, decompose non-enzymatically to a variety of strongly flavored primary and secondary compounds (Figure 1). This mechanism is different from that which occurs in animal and plant tissues. In animal tissue, oxidation of lipids occurs non-enzymatically, being initiated largely by hemo-proteins which then are decomposed enzymatically. In plants, lipid hydroperoxides are both enzymatically formed and enzymatically decomposed (17).

This brings us to the subject of lipid oxidation and soybean reversion flavor. As soybean becomes incorporated to a greater and greater degree in the human diet, the reversion flavor of soy assumes more significance. Many of the compounds attributed to the reversion flavor are products of lipid oxidation and are characterized as "beany, buttery, painty, fishy, grassy, or hay like" (5).

The linolenic acid component of soybean oil has been most frequently implicated in the formation of reversion flavor where it is present at about nine percent (18). Soybean oil also contains substantial amounts of oleic and linoleic acids as do cottonseed, corn and several other oils. These oils are not, however, subject to flavor reversion but then, they only contain less than one percent linolenic acid(18). It would seem that linolenic acid must occur in substantial amounts together with linoleic and possibly oleic acids in order for reversion products to occur. A satisfactory explanation of this has not yet been developed- and yet more than seventy compounds have been identified in the volatile fractions of reverted

LIPID PRECURSOR →		INTERMEDIATES →		DERIVED LIPIDS
Mono-,Di-,Tri-Glycerides	O_2	Hydroperoxides		Sat. & Unsat:
Free fatty acids	Autoxidation	Radicals	Autoxidation	acids
	Enzymolysis (lipoxigenase)		Enzymolysis (reductase)	aldehydes
				alcohols
	Catalysis by Metallo-compds. (free metals; heme compds.)		Catalysis by Metallo-compds. (free metals; heme compds.)	ketones
				esters
	Irradiation		Irradiation	lactones
	Light		Light	other ring structures
	Heat		Heat	hydrocarbons

Figure 1. General scheme for lipid degradation

soybean. These include methyl ketones, esters, saturated and unsaturated aldehydes, and acids. These compounds can be formed in different ways including autoxidation or by way of lipoxygenases and other enzymes to form hydroperoxides. The breakdown of hydroperoxides results in the formation of hexanal, hexanol, 2-hexenal, ethyl vinyl ketone, and 2-pentyl furan, all of which are characterized as "beany", or "grassy" (5). The 2-pentyl furan is particularly noteworthy because at levels as low as 1-10 ppm it produces a beany flavor while at higher concentration, it assumes a licorice-like character (19).

The biogenesis of flavors in plants also involves lipid oxidation. Fruits and vegetables have volatile compounds that are genetically controlled which are responsible for their flavor. These compounds are formed during the maturation and post-harvest storage through specific enzymatic changes in the mono- and disaccharides, in the amino acids, and in certain unsaturated lipids that contain the 1,4-pentadiene structure. These lipids, mainly the linoleic and linolenic acids, are oxidized to their hydroperoxides by action of specific lipoxygenases which in turn undergo other enzymatic transformations yeilding specific aldehydes and other secondary compounds. The presence of aliphatic aldehydes is considered an important occurance because they are not only aromatically potent but are generally unstable. These aldehydes, together with their corresponding alcohols, are responsible for many of the characteristic flavors in food plants including banana, apple, peas, plums and grapes. Their fresh green character is largely due to hexanal, and 2-hexenal which derive from linoleic and linolenic acids. The character of cucumber also derived from these lipids is mainly due to 2-nonenal and to 2,6-nonadienal and their corresponding alcohols (15). In tomatoes, the fresh aroma is due largely to cis-3-hexenal, cis-3-hexenol, and trans-2-hexenal which derive from linolenic acid (20). The principle flavorant of mushroom is 1-octen-3-ol which is derived from linoleic acid (21). The character of green beans is in part due to hexanal, 2-hexenal, and 1-octen-3-ol derived from linolenic together with linoleic acid (12). Characteristic pea flavor is due to a combination of C3,5,6 saturated alcohols, C7,8,9 2-enals, C9,10 2,4-dienals as well as 2-pentyl furan which are all derived from linoleic acid (13).

Beverage milk, an excellent vehicle for demonstrating the nature of lipid flavors is an oil in water emulsion. It has a delicate yet complex flavor.

When fresh, it is acceptable to most people who might otherwise reject it out-of-hand if rancid or tainted flavors were detected. The normal compliment of free fatty acids in milk originated through action of native lipase which hydrolyzes milk fat. These acids are located mainly in the fat globules from which they came and where their flavors are largely masked, and together with short chain free fatty acids found in the aqueous portion, provide milk with its normal flavor. If the fatty acids located in the fat globules could be shifted to the serum, the resulting milk would be considered unacceptably rancid. Small additions of fatty acids to the aqueous phase are usually tolerated but once past the threshold of rancidity, the resulting milk may no longer be acceptable. To be sure, in some areas of the world where refrigeration is generally unavailable, a higher degree of rancidity is tolerated or even preferred. In contrast, it takes only trace quantities of certain weeds such as wild onions in the cow's feed supply to taint the flavor of milk. It was reported that as little as two milligrams of 2,6-nonadienal spoils the flavor of one ton of fat (1).

Another product which depends on lipids for most of its flavor is blue mold cheese. Here, the flavor is developed through selective lipolysis of milk fat yeilding mainly small chain saturated fatty acids. During the ripening period, some of these acids undergo enzymatic β-oxidation, decarboxylation, and reduction to yeild a mixture of fatty acids, methyl ketones and methyl carbinols. Depending upon their polarity, these compounds are partitioned between the fat and the aqueous phases of the cheese and, as in the case of milk, the partitioning ratio was found to be critical to the normal flavor (22).

Some of the most appetizing aromas recognized throughout the world are associated with certain food products that have something in common, that being that each had received heat at some point during processing. Examples of these are the aromas eminating from hot, baked bread, roasted coffee and nuts, butterscotch, barbeque, roast beef, pork, and poultry, and others. Compounds responsible for these delicious aromas are derived from non-enzymatic browning of sugar, aided by amino acids, and by products of lipid oxidation. For several years, flavor manufacturers have recognized the commercial value of those flavors that evolve when mixtures of reducing sugars, amines and lipids are heated together. They are searching for ideal conditions that recreate synthetically

TABLE 1

ALIPHATIC ALDEHYDES

General Contribution to Flavor:

Saturated- power, warmth, resonance, depth, roundness, freshness
2-Enals & 2,4-Dienals- sweet, fruity to fatty and oily

Homologous Series:

	Sat.	Flavor, (+)(-)	2-Enals	Flavor, (+)(-)	2,4-Dienals	Flavor, (+)(-)
C_2	fresh, pungent	(+)C_2 ubiquitous				2,6-nonadienal:
				(+)cis-4-heptenal: butter, cream (1)		(-)linseed oil
C_3	fresh, milky	(+)C_{1-18} pork (23)				(+)cucumber (32),(15)
C_4	↓	(+)C_{2-16} beef (24)	sweet, pungent	(+)$C_{4,6-12}$ ham (27)		(-)$C_{7,10}$ oxidized milk (1)
C_5	↓	(+)C_{2-5} coffee, cocoa	sweet, green	(-)C_{4-11} skim milk (28)		
C_6	fresh, green	(+)C_6 fruit	↓	(-)C_{5-8} soy & veg. oils (25)		(-)$C_{7,10}$ soy & oxidized veg. oil (30)
C_7	↓		↓		sweet, oily	
C_8	fresh, citrus	(-)C_6 veg. oils, skim milk (25)	↓	(-)C_{5-10} oxidized milk (1)	↓	
C_9	↓	(+)C_{3-9} gr. veg. (26)	sweet, fatty, green		↓	(+)$C_{7,9-12}$ ham (23)
C_{10}	↓	(-)C_{5-10} oxidized milk (1)	↓	(+)C_{6-13} beef (29)	↓	(+)C_{8-12} beef (29)
C_{11}	fatty		sweet, fatty	(+)C_6 fruit, banana	↓	
C_{12}	↓		↓	(+)$C_{6,9}$ cucumber (15)	↓	(+)C_{8-12} (31) chocolate

TABLE 2

ALIPHATIC ALCOHOLS

General Contribution to Flavor: similar to n-aldehydes but milder

Homologous Series			
C_2	no comment	(-) C_2-C_{12}	heated vegetable oils (7)
C_3	solventy, non-descript	(+)$C_{4,7}$	fish and shellfish (33)
C_4	↓	(-)C_2-C_9	oxidized milk fat (1)
C_5	↓		
C_6	grassy green		
C_7	↓		
C_8	↓		
C_9	fatty, green		
C_{10}	fatty		
C_{11}	↓		

these naturally occuring flavors. We can find products in the market that reflect this activity such as artificially flavored analogs of cheeses, beef, pork, bacon and poultry.

In order to understand better the contribution of derived lipids to the flavor of food, it seemed logical to characterize them organoleptically. This was done by a team of flavor chemists from Stepan Flavors and Fragrances who undertook to characterize the flavor and odor of six homologous series of aliphatic compounds namely the n-aldehydes (sat., 2-enals, 2,4-dienals), n-acids, methyl ketones, methyl carbinols, n-alcohols, ethyl esters, and gamma and delta lactones. (Figure 2 illustrates simplified routes of formation of these compounds.) This exercise revealed the following facts about the nature of aliphatic compounds as they relate to flavor:
1) Each class had an overriding character that was unique and readily identifiable.
2) Close members within each series were more similar to each other than to distant members.
3) Carbonyls and alcohols of the same chain length and the same oxygen position were found similar in character. For example methyl ketones were similar to methyl carbinols; ethyl ketones were similar to ethyl carbinols and n-aldehydes were similar to n-alcohols.

These characterizations are presented in Tables 3-8. An inspection of the odor character of the members of the various groups of compounds generally gives a good indication as to whether a compound will have a positive or negative effect on the overall flavor of a particular food product. For example, the greenness of 2-hexenal may be an integral part of a fresh fruit or vegetable flavor but would be out of place in milk. If it was indicated in the literature that a compound was a part of the normal flavor of a food, its effect was assumed to be positive. Compounds appearing as a result of rancidity or oxidative changes in a food, were assumed to have a negative effect on the flavor. Compounds identified in various food items and their positive or negative flavor contribution are characterized in Tables 3-8.

In summary, lipids have a profound influence on our interpretation of the food product in which they are found. Their contribution may be tactile as in the mouthfeel of ice cream. They may act as heating or masking agents and as resevoirs for fat soluble compounds.

Several pathways leading to derived lipids

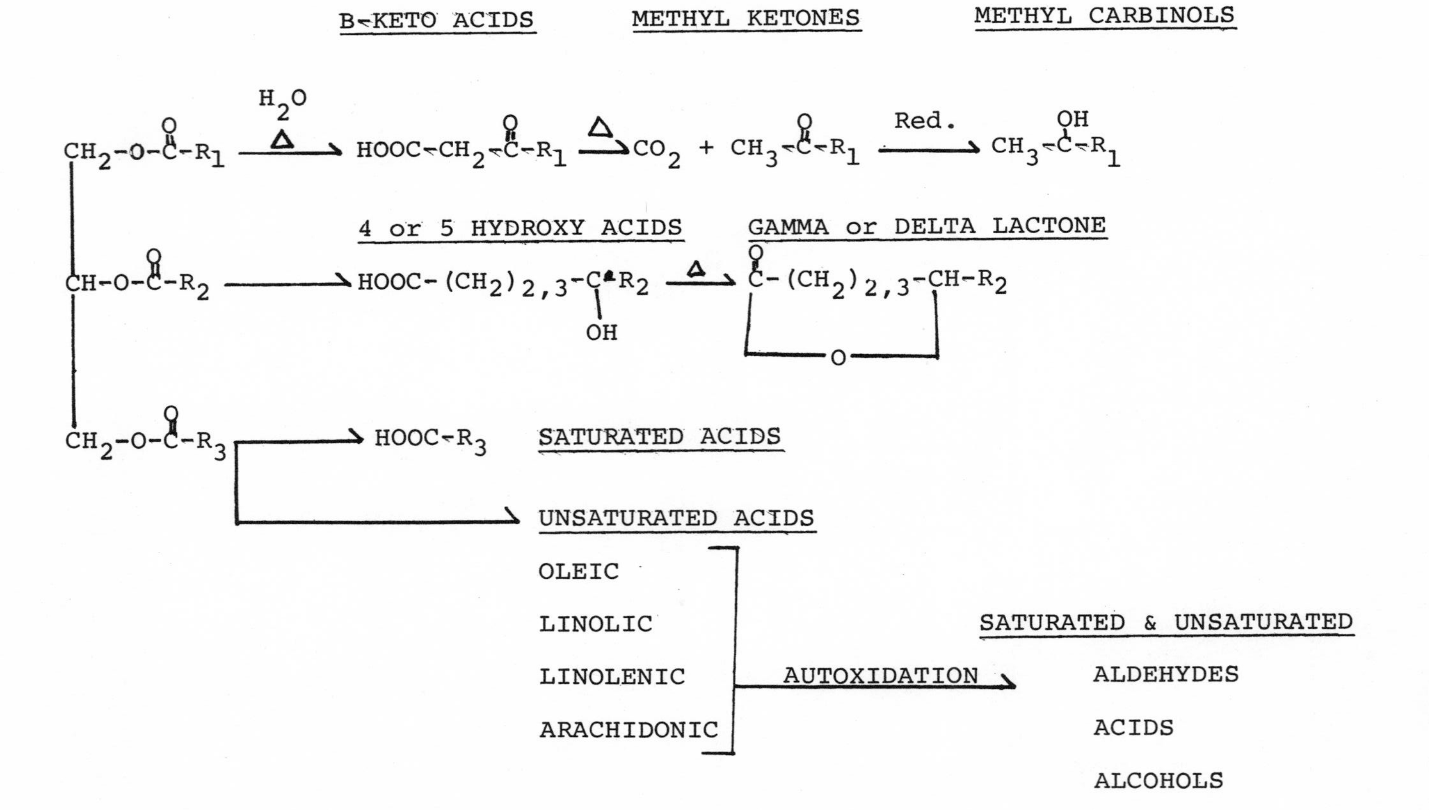

Figure 2. Degradation of lipids to important flavoring compounds

TABLE 3

METHYL KETONES

General Contribution to Flavor: adds a piercing, sweet fruitiness

Homologous Series		
C_3	pungent, sweet	(+)$C_{5,7}$apple
C_4	solvent, sweet	(+)$C_{5,7}$grape
C_5	↓	(+)$C_{5,7,9,11}$Blue cheese (22)
C_7	blue cheesey	(+)$C_{3,4,7-15}$beef, ham (34)
C_8	↓	(+)$C_{4,5,7,9,11}$butter (10)
C_9		(−)same, overheated milk (35)
C_{11}	fatty, sweet	(+)C_7mushroom (11)
		(−)C_7potato flakes (4)
		(−)ethyl vinyl ketone in soy oil (36)
		(+)3-octen-2-one in mushroom (11)

METHYL CARBINOLS

General Contribution to Flavor: parallels the methyl ketones but are less fruity, more grassy green

Homologous Series		
C_5	grassy, solvent	(+)supplements corresponding ketones
C_7	grassy, blue cheese	(−)C_5 soybean (37)
C_8	↓	(−)1-penten-3-ol, oily and grassy flavor in meat and milk (38)
C_9		(+)1-octen-3-ol in mushroom (39)
		(−)1-octen-3-ol butter; soy and other vegetable oils (39)

TABLE 4

ALIPHATIC ACIDS

General Contribution to Flavor: sour, fruity, cheesey and animal-like character

Homologous Series		
C_2	vinegary, sour	(+)$C_{2,4,6,8}$ tangy cheeses (40)
C_3	sour, swiss cheesey	(+)C_{1-16} white bread (41)
C_4	sweaty, cheesey	(-)C_{1-5} spoiling wet fish (3)
C_5	↓	(+)C_{1-6} beef (42)
C_6	↓	$C_{4,6,8,10,12,14}$ (±) background flavor of milk
C_7	↓	(-) when too concentrated in milk (34)
C_8	goaty, cheesey	(-) 2-enoic C_{6-12} in oxidized corn oil (2)
C_9	parafinic	(-) 2-enoic $C_{8,9,10,12}$ heated pork fat (43)
C_{10}	↓	(+) succinic acid - meaty note in shellfish (44)
C_{12}	↓	
C_{14-18}	very little odor	

TABLE 5

LACTONES

General Contribution to Flavor: adds rich, creamy, fruity, deep, full flavor

Gamma Homologous Series			Delta Homologous Series	
C_4	oily	(+)$C_{4,5,8}$ roasted nuts (45)		(+)C_{6-18} heated milk, commercial magarine (49)
C_5	creamy, tobacco	(+)C_{4-10} beef (29)		
C_6	creamy, coconut	(+)$C_{4,6,8,9}$ tomato (14)	very similar to gamma series but smoother, less fruity, more dairylike	(-)C_{6-18} milk when too high; whole dried milk, stored butter (1)
C_7	↓	(+)C_{5-8} fried corn oil (19)		
C_8	coconut	(+)C_7 pineapple (46)		
C_9	↓	(-)C_7 oxidized cottonseed and soy oil (47)		(+)$C_{5,7,8,9}$ beef (29)
C_{10}	peachy	(+)C_{10} strawberry (48)		(+)$C_{8,10,12,14,16}$ heated animal fat (50)
C_{11}	↓	(+)$C_{10,16}$ heated milk fat (9)		(+)$C_{8,12,10,14}$ coconut oil (51)
C_{12}	↓			(+)$C_{9,10}$ pork (35)
				(+)$C_{10,12}$ peaches apricots (47)

TABLE 6

ALIPHATIC ETHYL ESTERS

General Contribution to Flavor: sweet, fruity, pleasant character

Homologous Series		
C_4	sweet, solventy	(+) in all fruits and berries and in all fermented products (52)
C_5	rummy	(-) in dairy products if concentration is too high (1)
C_6	fruity	(-) ethyl butyrate and hexanoate in cheddar cheese (53)
C_7	apple	(+) 2,4 (or 6) alkadienoates in pears (54)
C_8	pineapple	(+) amyl and butyl acetate in banana (52)
C_9	pear	
C_{10}	orange	
C_{11}	wine	
C_{12}	cognac	
C_{13}	sweet, fatty	
C_{14}	sweet, fatty	

include autoxidation, enzymolysis, and catalysis. The derived products may contribute positively or negatively to the flavor of foods. There is nothing as delightful as a lipid in its proper place or so noxious as one out of context.

Literature Cited

1) Kinsella, J. Proceedings, Frontiers in Food Research, (1968), p.94.
2) Kawada, T., Krishnamurthy, R., Mookherjee, B., and S. Chang. J. Am. Oil Chem. Soc., (1967), 44, P.131.
3) Jones, N. "In Symposium on Foods: The Chemistry and Phisiology of Flavors", edited by H. Schultz, E. Day, L. Libbey, p.267, Avi Publ. Co., Westport, Conn., (1967).
4) Sapers, G., Panasiuk, O., Talley, F., Osman, S., and R. Shaw, J. Fd. Sci., (1972), 37, p.579.
5) Fennema, O., "Principles of Food Science Part I - Food Chemistry", Marcel Dekker, Inc., New York, (1976).
6) Wilson, B., Pearson, A., and F. Shorland. J. Agr. Fd. Chem., (1976), 24, (1), p.7.
7) May, W., Ph. D. Thesis Submitted to Rutgers, The State University, June, (1971).
8) van der Wal, B., Kettenes, D. Staffelsma, J., Sipma, G., and A. Semper. J. Agr. Fd. Chem., (1971), 19, p.276.
9) Forss, D., Urbach, G., and W. Stark, Intern. Dairy Congr. Prac. 17th, (1966), C2, p.211.
10)Winter, M., Malet, G., Pfeiffer, M., and E. Demole. Helv. Chim. Acta, (1962), 45, p.1250.
11) Thomas, A. J. Agr. Fd. Chem., (1973), 21, (6), p.955.
12)Stone, E., Hall, R., Forsythe, R. and Kazeniac, S. Unpubl. to date, presented at Am. Chem. Soc., Fall, (1976).
13)Whitfield, F., and J. Shipton. J. Fd. Res., (1966), 31, p.328.
14)Viani, R., Bricout, J., Marion, J., Muggler - Chavan, F., Reymond, D., and R. Egli. Helv. Chim. Acta, (1969), 52, p.887.
15)Forss, D., Dunstone, E., Ramshaw, E., and W. Stark. J. Fd. Sci., (1962), 27, p.90.
16)Schultz, H., Day, E., and L. Libbey. "Symposium on Foods: The Chemistry and Physiology of Flavors", Avi Publ. Co., Westport, Conn. (1967).
17)Eriksson, C. J. Agri. Fd. Chem., (1975), 23, (2), p.126.
18)Sherwin, E., J. Am. Oil Chemists' Soc., (1976), 53, p.430.
19)Krishnamurthy, R., Smouse, T., Mookerjee, B., Reddy, B., and S. Chang. J. Fd. Sci., (1967), 32, p.372.

20)Stone, E., Hall, R., and S. Kazeneac. J. Fd. Sci,, (1975), 40, p.1138.
21)Stone, E., Hall, R., and S. Kazeniac. Presented IFT convention, Philidelphia, Penn., June, (1977).
22)Kinsella, J., and D. Hwang. Critical Reviews in Food Science and Nutrition, Nov., (1976), p.191.
23)Lillard, D., and J. Ayres. Food Tech., (1969), 23, (2), p.251.
24)Herz, K. Ph. D. Thesis, Rutgers - The State University, (1968).
25)Kawahara, F., and H. Dutton. J. Am. Oil Chem. soc., (1952), 29, p.372.
26)Wilkinson, R. Internal Report #4, Division of Dairy Research, C.S.I.R.O., Melbourne, Australia, (1964).
28)Forss, D., Pont, E., and W. Stark. J. Dairy Research, (1955), 22, p.91.
29)Liebich, H., Douglas, D., Zlatkis, A., Muggler - Chavan, F. and A. Donzel. J. Agr. Fd. Chem., (1972), 20, p.96.
30)Hoffman, G. J. Am. Oil Chem. Soc., (1961), 38,p.31.
31)Boyd, E., Keeney, P., and S. Pattan. J. Fd. Sci., (1965), 30, p.854.
32)Keppler, J., Schols, J., Feenstra, W., and P. Meyboom. J. Am. Oil Chem. Soc., (1965), 42, p.246.
33)Ismail, R. Ph. D. Thesis, Louisiana State University and Agricultural and Mechanical College, (1971).
34)Mussinan, C., and J. Walradt. J. Agr. Fd. Chem., (1974), 22, p.827.
35)Forss, D. J. Agr. Fd. Chem., (1969), 17, (4),p.681.
36)Hill, F., and E. Hammond. J. Am. Oil Chem. Soc., (1965), 42, p.1148.
37)Arai, S., Koyanagi, O., and M. Fujimaki. Agr. Biol. Chem., (1967), 31, p.868.
38)Wilson, R., and I. Katz. J. Agr. Fd. Chem., (1972), 20, p.741.
39)Hoffman, G. J. Am. Oil Chem. Soc., (1962), 39, p.439.
40)Forss, D., and S. Pattan. J. Dairy Sci., (1966), 49, p.89.
41)Coffman, J. "In Symposium on Foods: The Chemistry and Physiology of Flavors", edited by H. Schultz, E. Day, L. Libbey, p.182, Avi Publ. Co., Westport Conn., (1967)
42)Yueh, M., and F. Strong. J. Agric. Fd. Chem., (1960), 8, p.491.
43)Watanabe, K., and Y. Sats. Agr. Biol. Chem., (1969), 33, p.1411.
44)Hashimoto, Y., "F.A.O. Symposium on the Significance of Fundamental Research in Utilization of Fish."

Husum, Germany. Paper #WP/II/6, (1964).
45)Walradt, J., Pittet, A., Kinlin, T., Muralidhaia, R., and A. Sanderson. J. Agr. Fd. Chem., (1971), 19, p.972.
46)Silverstein, R., Radin, J., Himel, C., and R. Leeper. J. Fd. Sci., (1965), 30, p.668.
47)Jennings, W. and M. Sevenants. J. Fd. Sci.,(1964), 29, p.158.
48)Willhalm. B., Palluy, E., and M. Winter. Helv. Chim. Acta., (1966), 49, p.65.
49)Calvert, L., and M. Przyblyska. Appl. Spectrosc. (1961), 15, p.39.
50)Boldingh, J., and R. Raylor. Nature (London), (1962), 194, p.909.
51)Allen, R. Chem. Ind. (London), (1965), p.1560.
52)Forss, D. Prog. Chem. Fats Other Lipids, (1973), 13, p.177.
53)Bills, D., Morgan, M., Libbey, L., and E. Day. J. Dairy Sci, (1965), 48, p.1168.
54)Jennings, W. "In Symposium on Foods: The Chemistry and Phisiology of Flavors", edited by H. Schultz, E. Day, and L. Libbey, p.419, Avi Publ. Co., Westport, Conn., (1967).

RECEIVED December 22, 1977

2

Chemistry of Deep Fat Fried Flavor

STEPHEN S. CHANG, ROBERT J. PETERSON, and CHI-TANG HO

Department of Food Science, Cook College, Rutgers, The State University of New Jersey, New Brunswick, NJ 08903

Deep-fat frying is one of the most commonly used procedures for the manufacture and preparation of foods in the world. The fast food restaurants which have been growing rapidly in recent years further increase the consumption of fried foods, especially fried chicken, fish and chips, and french fries. Evidently, a major portion of the 10 billion pounds of fats and oils consumed by Americans each year are used in fried foods. For example an estimated 500 million pounds of fats and oils are used each year for the manufacture of potato chips alone in the United States; another 200 million pounds are used each year for doughnuts; and 400 million pounds for frozen french fries.

Not long ago, we asked the sales manager of a large, international flavor company, what flavor he would like to see us develop which he believed would have a large market. Without hesitation, his answer was "deep-fat fried flavor".

In deep-fat frying, foods which usually contain moisture are continuously or repeatedly dipped into an oil which is heated to a high temperature of usually 185°C. in the presence of air. Under such conditions, both thermal and oxidative decomposition of the triglycerides may take place. Such unavoidable chemical reactions cause formation of both volatile and nonvolatile decomposition products.

The nonvolatile decomposition products contain oxidized and polymerized glycerides. They do not contribute to the desired deep-fat fried flavor of foods, except that they could make the food greasy with a possibly bitter taste. The deep-fat fried flavor is essentially contributed by the volatile decompositon products (VDP).

The systematic chemical identification of these VDP is important in at least three aspects. First, the mechanisms of the formation of these compounds may lead us to an understanding of the chemical reactions which take place during deep-fat frying.

0-8412-0418-7/78/47-075-018$10.00/0

Second, the VDP are inhaled by the operators of deep-fat frying, particularly, restaurant cooks. Furthermore, it has been shown by our investigation that a portion of the VDP remains in the frying oil, thus entering the consumer's diet. An understanding of their chemical identities may facilitate the investigation of their effect upon human health.

Lastly, the flavor of deep-fat fried foods is partly due to the VDP. A knowledge of their chemical composition may make possible the manufacture of a synthetic flavor which can be used to enhance the flavor of deep-fat fried foods, or to manufacture foods with a deep-fat fried flavor without the necessity of the frying process.

Volatile Flavor Constituents (VFC) in Deep-fat Fried Foods

The VFC in deep-fat fried foods can be originated either from the oil, or from the food, or from the interaction between the oil and the food. When six different oils and one fat, as shown in Table I were used for simulated deep-fat frying, using moist cotton balls to replace the food, each oil developed a different flavor after 20 fryings in six hrs. at 185 ± 5°C. (1). The moist cotton balls containing 75% of water by weight are similar to inert pieces of potato. However, they do not contribute any flavor to the frying oil. This simulated deep-fat frying avoided the use of food, the odors and flavors of which might be so pronounced as to make the study of the odor and flavor originating from the oil itself most difficult, if not impossible.

This design, of course, assumes that the flavor of the frying oil would contribute to the total flavor of fried foods. It is obvious that if the oil, after being used for frying, has a strong, pleasant flavor, then it will certainly enhance the desirability of the fried food. On the other hand, if the oil, after being used for frying, develops a strong, unpleasant flavor, it would make the fried food less desirable. While this assumption has no experimental data to support it, it nevertheless appears logical and is therefore used as a preliminary step to approach a complicated problem.

The six used frying oils and one fat were then organoleptically evaluated by a panel of 10, trained, experienced members for odor strength, odor pleasantness, flavor strength, and flavor pleasantness. A Hedonic scale of 1-9 was used, using 1 for the weakest odor and flavor, 9 for the strongest odor or flavor, 1 for the least liked odor or flavor and 9 for the most liked odor and flavor. Therefore, the higher the score indicated the stronger the odor or more pleasant the odor. The oils were arranged in increasing order of the strength of either their odor or flavor, as shown in Table II. They are also listed in decreased order of their pleasantness. For example, corn oil has the lowest strength

TABLE I

Oils and Fat Used for Simulated Deep-fat Frying

Oil or Fat	Iodine Value (Wijs)
Corn Oil	127
Cottonseed Oil	110
Peanut Oil	96
Soybean Oil, Hydrogenated and Winterized	115
Soybean Oil, Hydrogenated and Winterized	89
Soybean Oil, Hydrogenated	70

TABLE II

Relative Ranking of Oils by Adjusted Organoleptic Score

	Strength			Pleasantness	
	Odor	Flavor		Odor	Flavor
Corn	3.78	4.23	Corn	5.60	5.58
Peanut	3.73	4.63	Cottonseed	5.30	4.96
Cottonseed	5.08	5.03	Peanut	4.45	4.93
Soybean IV[a] 89	5.33	5.08	Soybean IV 89	4.35	4.48
Soybean IV 115	6.53	6.53	Soybean IV 70	3.30	3.03
Soybean IV 70	6.88	7.88	Soybean IV 115	3.00	2.48
Tukey (0.05) q =	2.44	2.40	Tukey (0.05) q =	2.33	2.21

[a]IV = Iodine Value

but the highest desirability, both in odor and in flavor, while the hydrogenated and winterized soybean oil with an IV of 115 had almost the highest strength, but the most undesirable odor and flavor.

Generally, the unhydrogenated oils were ranked higher than the hydrogenated soybean oils. Among the three hydrogenated oils, the one with an IV of 89 ranked the highest. This might be due to the fact that its oxidation is stabilized by hydrogenation. At the same time, there was not too much hydrogenation to yield relatively large amounts of hydrogenation flavor.

The VFC isolated from each of the six used frying oils and fat yielded gas chromatograms which were qualitatively and quantitatively different from each other (Figures 1 and 2). The gas chromatogram could, therefore, be used as a profile for the odor and flavor of the oil. The area of the 24 selected peaks common to all the profile curves and the average organoleptic panel scores of each of the oils were analyzed by computer, using a multiple stepwise regression method. Strong correlations between some peak areas (Table III) in the profile curves and the average organoleptic scores were found. The R^2 value for the correlation of each of the odor and flavor characteristics with one or two peaks, respectively, are shown in Table IV.

Volatile Flavor Constituents Originated from the Food Used for Frying

A systematic analysis of the VFC isolated from potato chips led to the identification of 53 compounds (2). Some of them, such as 3-*cis*-hexanal and 2,4-*trans*, *trans*-decadienal, were evidently produced by the frying oil. Some other compounds, such as dimethyl disulfide and 2,5-dimethyl pyrazine, were evidently produced by the potato.

Volatile Flavor Constituents Produced by the Interaction between the Food and the Oil

In order to study the mechanism for the formation of the VFC in potato chips from the constituents of the potato, cotton balls, moistened with water in which different constituents of the potato were dissolved, either individually or in various combinations, were fried in cottonseed oil. Five different amino acids were selected, according to their chemical structures (Table IV). Each of them was treated under deep-fat frying conditions with the use of moist cotton balls. The aroma thus generated by each of the amino acids, as judged by an experienced panel of six persons, is shown in Table IV.

The strong, potato chip-like odor could be produced by either D- or L- isomers of methionine or their mixtures. After

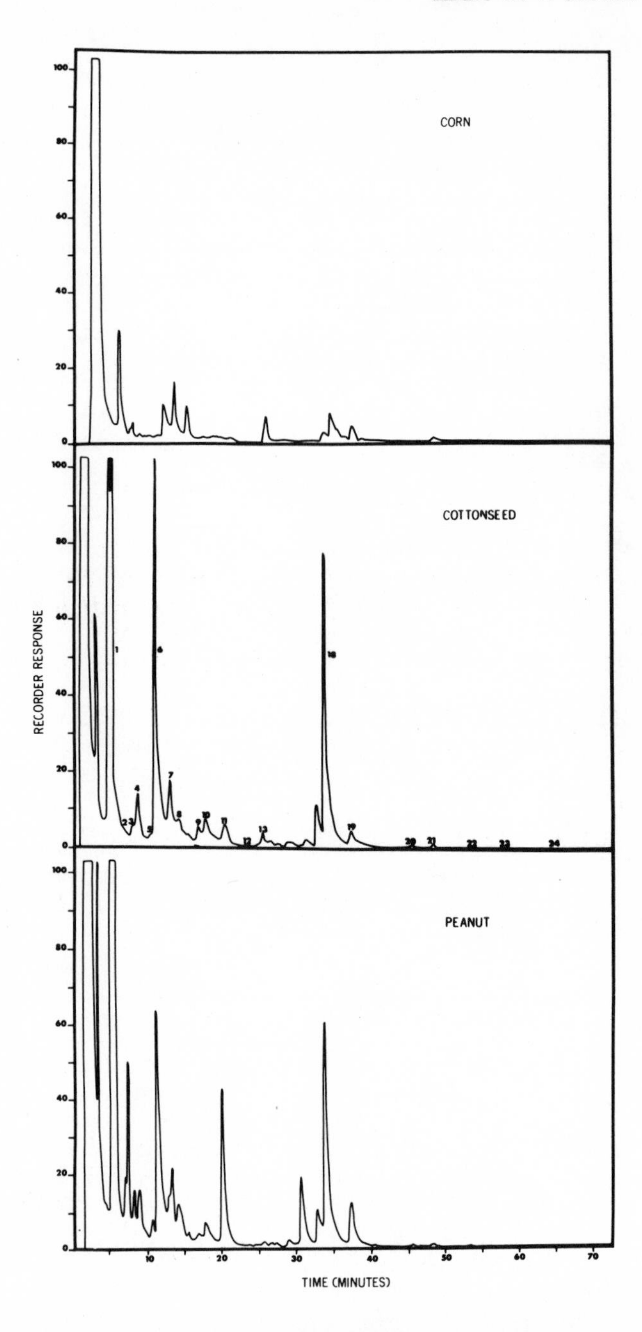

Figure 1. Gas chromatogram of volatiles isolated from used corn oil (top), *used cottonseed oil* (center), *and used peanut oil* (bottom)

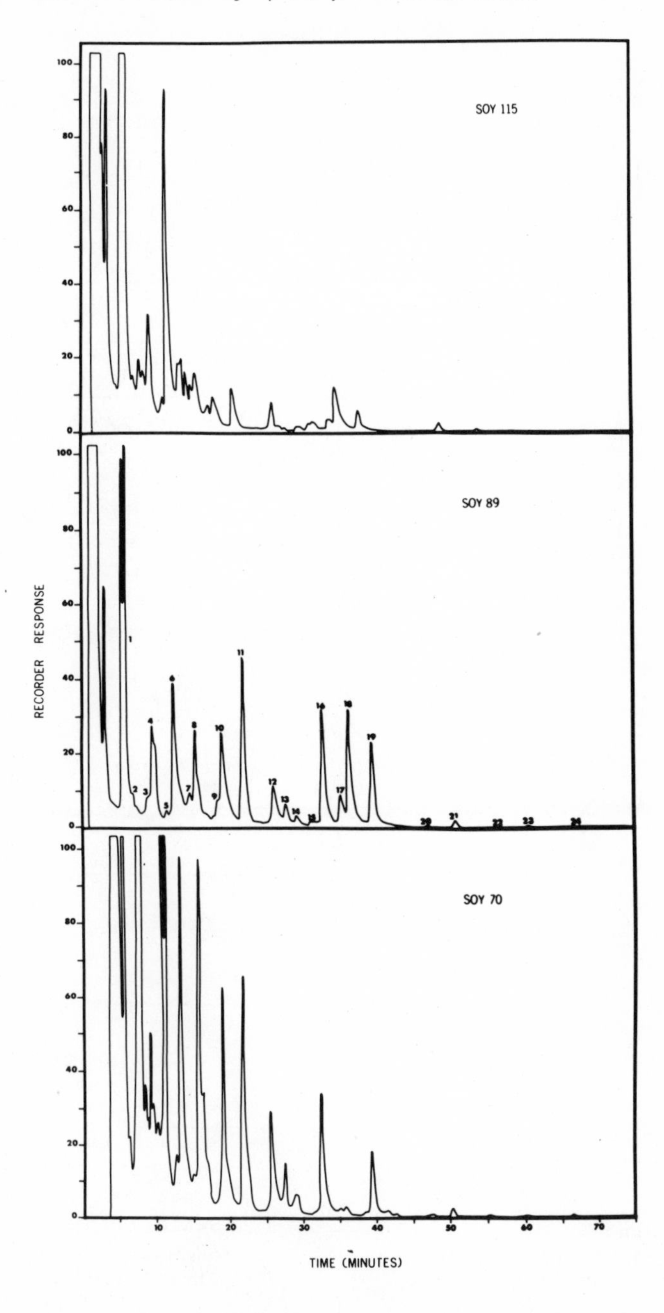

Figure 2. Gas chromatogram of volatiles isolated from used soybean oil with an iodine value (IV) of 115 (top), *IV 89* (center), *and IV 70* (bottom)

TABLE III

Summary of Stepwise Regression Analysis Used to Examine the Relationship between Peak Areas and Organoleptic Scores

	Correlated with peak numbers	R^2	
		1 peak	2 peaks
Stength of Odor	4 and 11	0.7909	0.9158
Pleasantness of Odor	8 and 4	0.9172	0.9882
Strength of Flavor	3 and 1	0.9303	0.9679
Pleasantness of Flavor	8 and 20	0.8778	0.9898

TABLE IV

Aroma Generated by Different Amino Acids Under Simulated Deep-fat Frying Conditions

Amino Acid	Aroma
Threonine	Wet, hair, earthy
Proline	Stale popcorn, bitter
Histidine	Stale popcorn
Cystine	Slightly meaty
Methionine	Strong potato chip-like

the cottonseed oil was used for the deep-fat frying of moist cotton balls containing methionine, it retained a strong, potato chip-like flavor (3).

The chemical structure required for the production of the potato chip-like flavor is quite specific as shown in Table V. The VFC in this oil can be isolated by subjecting the oil to 90°C under a vacuum of < 0.01 mm Hg. for 1½ hrs. The isolated volatile compounds had a strong, potato chip-like aroma. When they were dropped on a perfumer's paper stick, the aroma lingered on the stick for many hours, thus indicating that the compounds responsible for the potato chip-like aroma were of relatively high volume points.

The isolated VFC were fractionated by gas chromatography and the pure fractions were then identified by infrared and mass spectrometry. Among the compounds tentatively identified were 2-methyl mercaptomethyl butanal, 2-methyl mercaptosulfoxide-2-pentenal, 2-methylmercapto-5-methyl-2-hexenal, and 2-methylmercapto-2,4,6-octatrienal. These compounds are probably produced by the interaction between decomposition products of the amino acids and the decomposition products of oils at frying temperature.

Systematic Identification of the VDP of Frying

In order to elucidate the chemical structures of the VDP which were formed by the triglycerides during deep-fat frying, without the complication and interference of the food fried, an inert material must be found to substitute for the food. After trying with various materials, it was found that moist cotton balls containing 75% by weight of water were a good simulation of an inert piece of potato. They were used for simulated deep-fat frying in the apparatus as shown in Figure 3, which was designed to produce nonvolatile decomposition products, as well as to collect the VDP produced during deep-fat frying (4). The aluminum frying basket (A) was held in position by clamping at (B) and (C). The top of the Sunbeam deep-fat fryer was fitted with an Alembic-type cone (F) made of stainless steel. The cone was 10 in. high and had a top diameter of 5 in.; bottom diameter, 11 in. It was cooled with running water through aluminum coils wrapped around the outside of the cone. A glass connector (H) with an Alembic-shaped head (G) was used to join the condenser (I).

For frying, the deep-fat fryer containing 2300 ml of corn oil maintained at 185°C. was lowered until the aluminum basket was out of the oil. Ten moist cotton balls, each containing 75% by weight of water, were placed in the aluminum basket. A vacuum pump connected to the end of the flowmeter (P) was turned on to draw a current of air through the top of the fryer

TABLE V

Effect of Chemical Structure upon the Production of a Potato Chip-like Flavor by Methionine under Deep-fat Frying Conditions

Compounds	Structures	Characteristic of flavor produced under deep-fat frying conditions
D-Methionine L-Methionine DL-Methionine	CH_3-S-CH_2-CH_2-CH(NH_2)COOH	Good potato chip-like
S-Methyl-L-cysteine	CH_3-S-CH_2-CH(NH_2)COOH	Good potato chip-like
DL-Ethionine	CH_3-CH_2-S-CH_2-CH_2-CH(NH_2)COOH	Good potato chip-like
S-Ethyl-L-cysteine	CH_3-CH_2-S-CH_2-CH(NH_2)COOH	Obnoxious (cooked turnip)
Methionine hydroxy analog	CH_3-S-CH_2-CH_2-CH(OH)COOH	Obnoxious (cooked turnip)
S-Carboxymethyl-L-cysteine	HOOC-CH_2-S-CH_2-CH(NH_2)COOH	Obnoxious (cooked turnip)

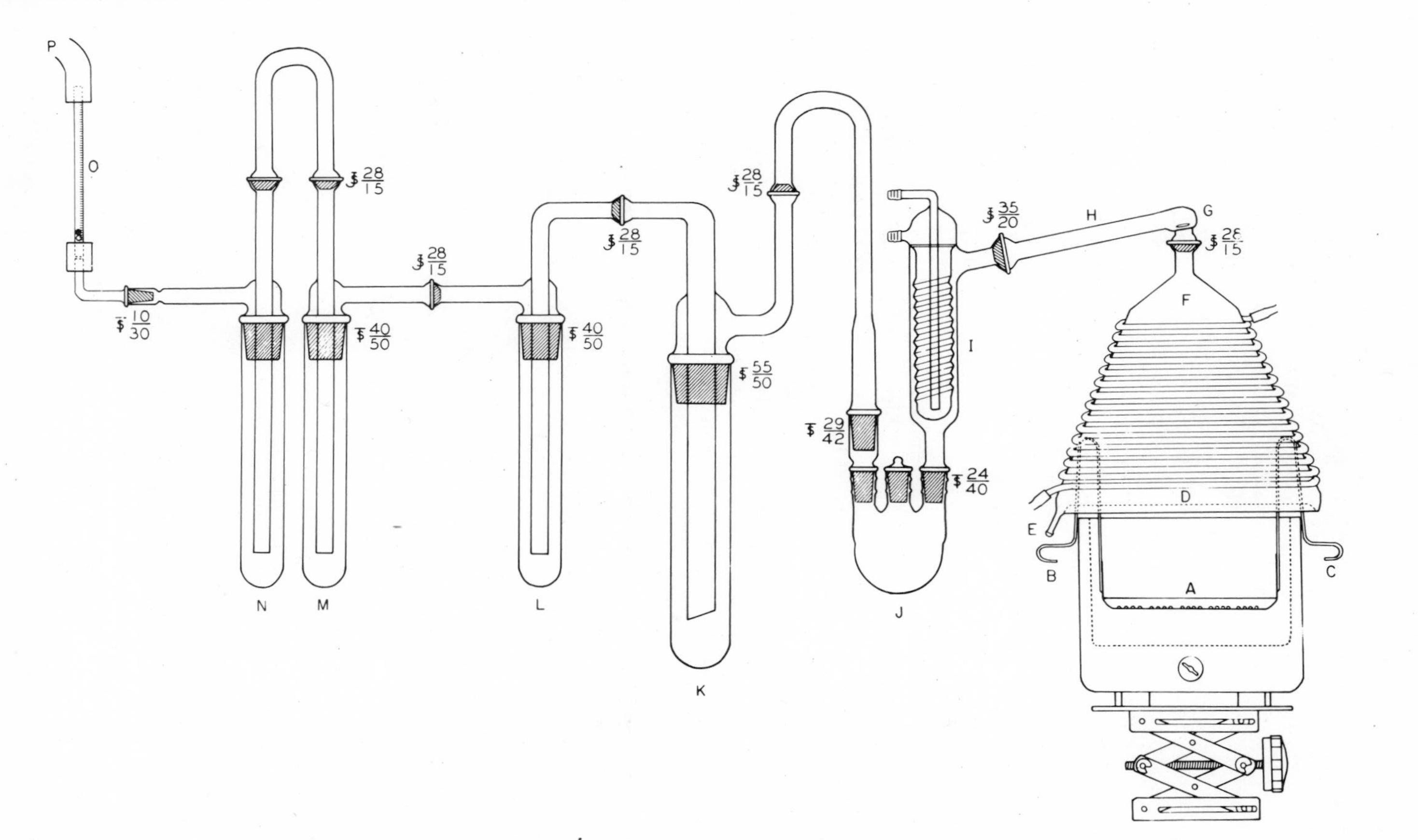

Figure 3. Laboratory apparatus for deep fat frying under simulated restaurant conditions

and then through the train of traps at a rate of 7.2 liters/min. as indicated by the flowmeter (O). The fryer was then raised until the cotton balls were immersed in the oil and fried.

The VDP and steam thus produced were drawn by the current of air flowing through the apparatus into the stainless steel cone and then the condenser and the train of traps (K - N). The condensate collected on the inside of the cone could not drip back into the fryer because it was trapped by the Alembic edge (D). Excessive amounts of VDP and steam condensed in the cone would flow out from the exit (E) and could be collected with a suitable container. Those which were not condensed in the cone were collected in the flask (J) and traps (K - N). Those condensed in the head of the connector tube (H) also could not drip back into the fryer because of the Alembic head (G).

Ten moist cotton balls containing approximately 16 g. of water were fried every 30 min. Thirteen frying operations were done each day in 6 hrs. After each 12 hrs. of frying, 800 ml. of fresh corn oil were added into the fryer to replenish the oil absorbed by the cotton balls. After each 6 hrs. of frying, the oil was allowed to cool to room temperature. The apparatus was disassembled and all the condensates were washed out with distilled water and ethyl ether.

Difficulty arose when pure synthetic triglycerides were used for simulated deep-fat frying so that the nonvolatile decomposition products (NVDP) could be more easily characterized than mixed triglycerides. The cotton balls would absorb too much of the expensive pure triglycerides and would make the experiment financially impossible. The apparatus was therefore modified as shown in Figure 4. so that steam could be periodically injected into the heated pure triglycerides to simulate frying (5).

The steam generator was constructed from a three-necked round bottom flask (A). In one neck, a reflux condenser (C) was connected through a large bore stopcock (B). The center neck was fitted with a long glass tubing (D) extending to the bottom of the flask. The third neck was connected to an aluminum tubing (H) through a flowmeter (E) and a three-way stopcock (F). The aluminum tubing was extended into the deep-fat fryer by soldering through the wall of the fryer. The section of aluminum tubing inside the fryer was perforated with pin holes at equidistance with the end closed and was bent to form a loop (J) lying on the bottom of the fryer. A heating tape (I) was wrapped around the connecting tube (G) to prevent condensation of steam.

With stopcock (B) open, the water in the round bottom flask was heated to a vigorous boil. To simulate frying, the stopcock (B) was closed. When the desired degree of steam pressure was built up in the flask, the three-way stopcock (F) was opened to allow the steam to bubble through the 2 Kg. of pure triglycerides maintained at 185°C. in the Sunbeam deep-fat fryer (K). To stop the steam flow, stopcock B was opened. The three-way stopcock (F)

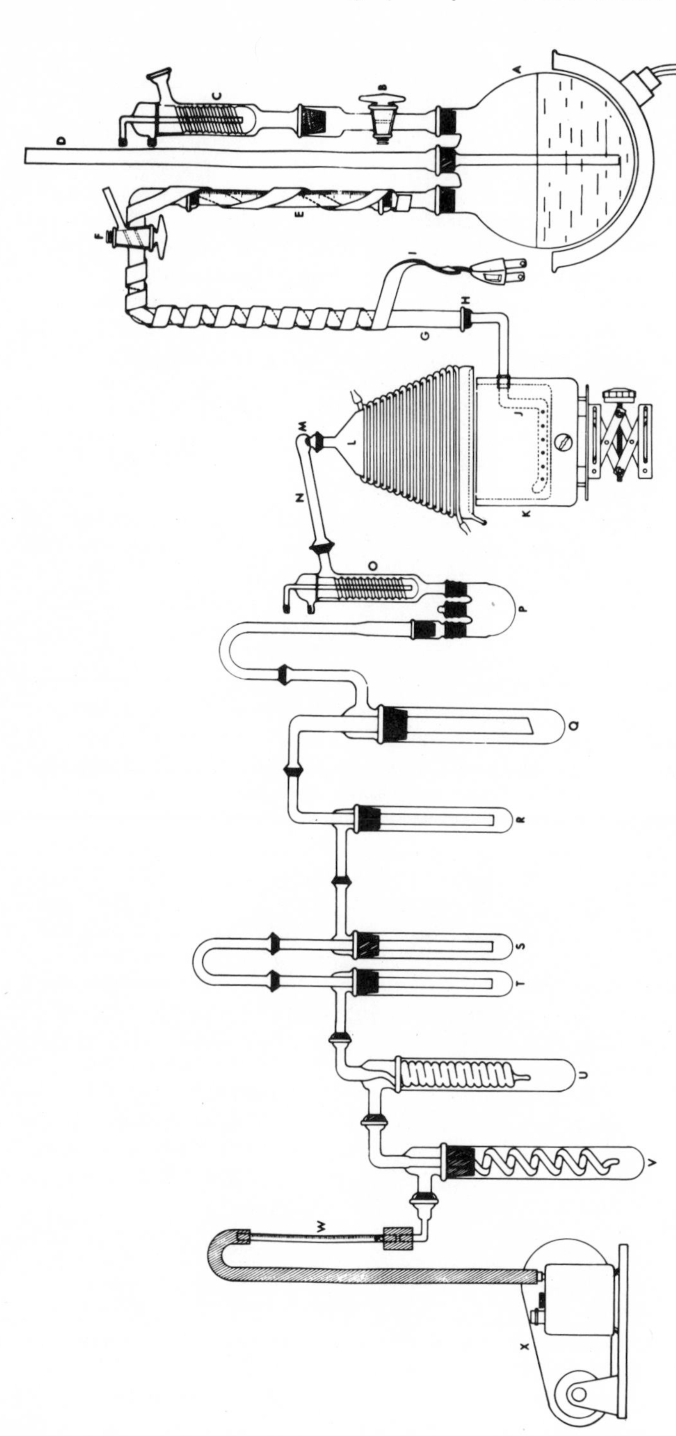

Figure 4. Apparatus used for treating pure triglycerides under simulated deep-fat frying conditions

TABLE VI

Volatile Decomposition Products Produced by Various Triglycerides During Simulated Deep-fat Frying

	Relative Amount of Compound			
Compound	Corn Oil	Hydrogenated Cottonseed Oil	Trilinolein	Triolein
I. Acidic VDP				
A. Saturated Acids				
Acetic	S	-	-	-
Propanoic	S	-	M	-
Butanoic	S	-	S	M
Pentanoic	L	S	M	L
Hexanoic	XL	S	XL	L
Heptanoic	L	S	L	M
Octanoic	L	S	M	XL
Nonanoic	L	S	M	XL
Decanoic	S	S	M	L
Undecanoic	S	XS	-	M
Dodecanoic	S	S	-	M
Tridecanoic	-	L	-	S
Tetradecanoic	-	M	-	S
Pentadecanoic	-	XS	-	-
Hexadecanoic	-	XL	-	-
Heptadecanoic	-	XS	-	-
Octadecanoic	-	L	-	-
B. Unsaturated Acids				
trans-2-Butenoic	-	-	S	-
trans-2-Pentenoic	-	-	L	-
trans-2-Hexenoic	S	-	-	-
trans-2-Heptenoic	-	S	XL	-
trans-2-Octenoic	M	S	S	M
trans-2-Nonenoic	M	XS	XL	M
trans-2-Decenoic	-	-	XL	M
trans-2-Undecenoic	S	-	-	L
trans-2-Dodecenoic	-	-	-	S
trans-2-Tridecenoic	-	-	-	S
cis-2-Heptenoic	-	-	S	-
cis-2-Nonenoic	-	-	L	-
cis-2-Decenoic	-	-	S	-

TABLE VI
(2)

Compound	Relative Amount of Compound			
	Corn Oil	Hydrogenated Cottonseed Oil	Trilinolein	Triolein
B. Unsaturated Acids				
trans-3-Pentenoic	-	-	M	-
trans-3-Nonenoic	-	-	L	-
trans-3-Decenoic	S	S	-	XS
cis-3-Heptenoic	-	-	S	-
cis-3-Octenoic	-	-	S	M
cis-3-Nonenoic	-	-	S	S
cis-3-Decenoic	L	M	S	XS
cis-3-Undecenoic	-	-	S(tent.)	-
cis-3-Dodecenoic	-	-	-	M
cis-4-Nonenoic	-	-	S(tent.)	-
5-Hexenoic	-	-	S	L
6-Heptenoic	-	XS	L	L
7-Octenoic	-	S	S	L
10-Undecenoic	-	XS	-	-
Palmitoleic	-	XS	-	-
Elaidic	-	S	-	M
Oleic	-	XL	-	-
Linoleic	-	L	-	-
Linolenic	-	XS	-	-
cis-2-trans-4-Octadienoic	-	-	M	S (tent.)
trans-2-cis-4-Decadienoic	-	-	M	-
trans-2-trans-4-Decadienoic	-	-	M	-
C. Hydroxy Acids				
3-Hydroxyhexanoic	S	-	S	-
2-Hydroxyheptanoic	S	-	M	-
2-Hydroxyoctanoic	-	-	S	-
3-Hydroxyoctanoic	-	-	S(tent.)	-
5-Hydroxyoctanoic	-	XS(tent.)	-	-
5-Hydroxydecanoic	-	XS(tent.)	-	-
10-Hydroxy-cis-8-Hexadecenoic	-	-	-	XS (tent.)
D. Aldehydo Acids				
Octanedioic acid semialdehyde	-	XS	-	S

TABLE VI
(3)

Compound	Relative Amount of Compound			
	Corn Oil	Hydrogenated Cottonseed Oil	Trilinolein	Triolein
D. Aldehydo Acids				
Nonanedioic acid semialdehyde	-	XS	-	-
Decanedioic acid semialdehyde	-	XS	-	-
Undecanedioic acid semialdehyde	-	XS	-	-
Tetradecanedioic acid semialdehyde	-	XS	-	-
E. Keto Acids				
4-Oxohexanoic	M(tent.)	-	-	-
4-Oxoheptanoic	S (tent.)	XS (tent.)	S(tent.)	-
4-Oxooctanoic	S(tent.)	-	-	-
4-Oxononanoic	-	XS (tent.)	-	-
4-Oxo-*trans*-2-Octenoic	-	-	-	L
4-Oxo-*trans*-2-Nonenoic	-	-	-	M
4-Oxo-*trans*-2-Undecenoic	-	-	-	S
4-Oxo-*cis*-2-Dodecenoic	-	-	-	XS (tent.)
F. Dibasic Acids				
Hexanedioic	S	-	S	-
Heptanedioic	S	XS	XS	-
Octanedioic	M	XS	S	S
Nonanedioic	L	XS	-	M
Decanedioic	-	XS	-	-
Undecanedioic	-	XS	-	-
4-Oxoheptanedioic	-	-	XS(tent.)	-
II. Nonacidic VDP				
A. Saturated Hydrocarbons				
Hexane	-	-	-	XS
Heptane	S	-	-	M
Octane	S	S	-	S
Nonane	M	-	S	XL

TABLE VI
(4)

Compound	Corn Oil	Hydrogenated Cottonseed Oil	Trilinolein	Triolein
	Relative Amount of Compound			
A. Saturated Hydrocarbons				
Decane	L	M	M	M
Undecane	M	M	-	S
Dodecane	S	S	-	L
Tridecane	S	XS	-	-
Tetradecane	S	S	S	-
Pentadecane	-	S	XS	-
Hexadecane	-	S	XS	-
Heptadecane	-	S	-	-
Octadecane	-	XS	-	-
B. Unsaturated Hydrocarbons				
1-Octene	-	-	S	M
1-Nonene	-	-	S	M
1-Decene	-	-	-	S
1-Undecene	-	XS	-	-
trans-2-Octene	-	S	S	-
cis-2-Octene	-	-	S	-
trans-Undecene	-	-	S	-
trans-Dodecene	M	XS	S	-
trans-Tridecene	XS	S	-	-
trans-Tetradecene	S	XS	-	-
trans-Hexadecene	-	S	-	-
trans-Heptadecene	-	S(tent.)	-	-
trans-1,3-Octadiene	-	-	S(tent.)	-
trans-1,3-Nonadiene	-	-	S(tent.)	-
trans, _trans_-Tetradecadiene	-	-	S(tent.)	-
trans, _cis_-Tetradecadiene	-	-	S(tent.)	-
C. Alcohols				
Ethanol	-	-	M	-
1-Propanol	-	-	-	L
1-Butanol	S	M	L	M
1-Pentanol	XL	L	XL	-
1-Hexanol	S	M	S	L
1-Heptanol	-	L	S	L
1-Octanol	XL	M	L	L

TABLE VI
(5)

Compound	Relative Amount of Compound			
	Corn Oil	Hydrogenated Cottonseed Oil	Trilinolein	Triolein
C. Alcohols				
1-Decanol	-	S	-	-
1-Undecanol	-	-	M	-
1-Dodecanol	-	-	S	-
2-Hexanol	-	XS(tent.)	-	-
2-Octanol	-	-	M	-
3-Octanol	XL	-	S	-
1-Penten-3-ol	L	-	-	-
1-Octen-3-ol	XL	L	XL	-
D. Saturated Aldehydes				
Propanal	-	-	-	L
Butanal	-	-	S	M
Pentanal	XL	M	XL	-
Hexanal	XL	L	XL	M
Heptanal	XL	L	XL	L
Octanal	M	XL	M	XL
Nonanal	XL	XL	S	XL
Decanal	M	M	M	M
Undecanal	-	S	-	L
Dodecanal	-	XS	XS	M
Tridecanal	-	XS	-	-
Tetradecanal	-	XS	-	-
Pentadecanal	-	XS	-	-
3,4,5-Trimethyl-heptanal	L (tent.)	-	M(tent.)	-
4-Methoxy-3,3-dimethylbutanal	S (tent.)	-	S(tent.)	-
E. Unsaturated Aldehydes				
trans-2-Hexenal	M	M	M	S
trans-2-Heptenal	XL	XL	XL	M
trans-2-Octenal	XL	XL	XL	M
trans-2-Nonenal	XL	XL	M	L
trans-2-Decenal	XL	XL	M	XL
trans-2-Undecenal	S	S	-	XL
cis-2-Heptenal	-	-	S	-
cis-2-Octenal	-	-	S	-
cis-2-Nonenal	-	-	XS	-
cis-3-Hexenal	-	-	M(tent.)	-

TABLE VI
(6)

Compound	Corn Oil	Hydrogenated Cottonseed Oil	Trilinolein	Triolein
	Relative Amount of Compound			
E. Unsaturated Aldehydes				
trans-4-Hexenal	S	-	S(tent.)	-
trans-3-Decenal	S	-	-	M
5-Hexenal	-	-	-	M
6-Heptenal	-	-	-	M
7-Octenal	-	-	-	L
5-Methyl-4-hexenal	S(tent.)	-	-	-
4-Oxo-trans-2-octenal	-	-	-	L (tent.)
trans-2-cis-4-Heptadienal	-	-	M	-
trans-2-cis-4-Nonadienal	S	L	M	M
trans-2-trans-4-Nonadienal	L	M	XL	-
trans-2-trans-6-Nonadienal	XL	-	-	-
trans-2-cis-4-Decadienal	S	L	XS	-
trans-2-trans-4-Decadienal	XL	L	XL	-
F. Ketones				
2-Heptanone	S	-	L	S
2-Octanone	S	-	M	-
2-Nonanone	-	XS	S	M
2-Decanone	S	S	-	L
2-Undecanone	-	-	M	-
2-Dodecanone	-	S	XS	-
3-Heptanone	-	-	S	S
3-Octanone	-	XS	S	S
3-Nonanone	-	-	S	S
3-Decanone	-	-	M	-
3-Dodecanone	-	-	XS(tent.)	-
4-Octanone	-	-	-	M
4-Undecanone	M	XS	-	-
4-Dodecanone	S	-	-	-
1-Octen-3-one	S(tent.)	-	-	-
2-Methyl-3-octen-5-one	S(tent.)	-	S(tent.)	-

TABLE VI
(7)

Compound	Relative Amount of Compound Corn Oil	Hydrogenated Cottonseed Oil	Trilinolein	Triolein
F. Ketones				
trans-3-Nonen-2-one	XL	-	S(tent.)	-
trans-3-Undecen-2-one	-	-	S(tent.)	-
Nonenone	-	XS(tent.)	-	XS (tent.)
Dodecenone	-	-	-	XS (tent.)
1-Methoxy-3-hexanone	M(tent.)	-	L(tent.)	-
G. Esters				
Ethyl acetate	XL	XL	XL	XL
Butyl acetate	-	S	S	-
Hexyl formate	-	-	XS	L
Ethyl hexanoate	S	-	-	-
Octyl formate	S	-	-	L
Methyl nonanoate	-	-	S	-
Ethyl octanoate	-	-	-	S
Methyl dodecanoate	-	-	XS	-
trans-2-Octenyl-formate	-	-	-	S
Ethyl-*cis*-2-dodecenoate	-	-	-	S (tent.)
H. Lactones				
4-Hydroxypentanoic	S	-	M	-
4-Hydroxyhexanoic	L	-	S	-
4-Hydroxyheptanoic	S	XS	-	XS
4-Hydroxyoctanoic	L	-	M	S
4-Hydroxynonanoic	-	S	S	M
4-Hydroxydecanoic	-	S	S	S
5-Hydroxyhexanoic	-	-	S(tent.)	-
5-Hydroxydecanoic	-	-	S	-
6-Hydroxyhexanoic	-	-	S	-
4-Hydroxy-2-hexenoic	-	-	M	XS (tent.)
4-Hydroxy-2-heptenoic	S	-	XS	-

TABLE VI
(8)

Compound	Relative Amount of Compound			
	Corn Oil	Hydrogenated Cottonseed Oil	Trilinolein	Triolein
H. Lactones				
4-Hydroxy-2-octenoic	-	-	XS	M
4-Hydroxy-2-nonenoic	L	-	XL	-
4-Hydroxy-2-decenoic	-	-	S	-
4-Hydroxy-3-octenoic	-	-	-	S (tent.)
4-Hydroxy-3-nonenoic	-	-	XL(tent.)	-
5-Hydroxy-2-nonenoic	-	-	M(tent.)	-
I. Aromatic Compounds				
Toluene	S	-	-	-
Butylbenzene	-	-	S	-
Isobutylbenzene	M	-	-	-
Hexylbenzene	S	-	S	-
Phenol	L	-	-	-
Benzaldehyde	S	XS	M	-
Acetophenone	S(tent.)	-	-	-
4-Phenylbutanal	M(tent.)	-	XS	-
5-Phenylpentanal	S(tent.)	-	S	-
J. Miscellaneous Compounds				
2-Pentylfuran	XL	L	XL	-
1,4-Dioxane	-	-	L	-

XS indicates extra small gas chromatographic peaks; S, small; M, medium; L, large, XL, extra large.

was added to avoid back suction. It was open to air at all times, except when steam was bubbled through the triglycerides. The fryer was connected to the train of cold traps in the same manner as described previously.

By turning on the vacuum pump (X), a current of air was drawn through the top of the fryer and the series of cold traps at a rate of 7.2 liters/min, as indicated by the flowmeter (W). The air flew over the surface of the triglycerides without bubbling through them. This simulated the conditions of a commercial deep-fat fryer placed underneath an efficient hood. Steam, equivalent to 15 ml. of water was bubbled through the triglycerides in 2 min. The operation was repeated at intervals of 30 min. Fifteen operations were performed each day in 7 hrs. After the last operation, the triglycerides were cooled to room temperature and allowed to stand overnight. Total time of simulated frying was 74 hrs.

The VDP produced by corn oil (6, 7), hydrogenated cottonseed oil (8,9), trilinolein (10), and triolein (11), respectively, under simulated deep-fat frying conditions, were collected, separated into acidic and nonacidic compounds, fractionated by repeated gas chromatography with columns of a polar and a nonpolar stationary phase, consecutively, and the pure gas chromatographic fractions were then identified by a combination of retention time and infrared and mass spectrometry. The identifications were finally confirmed with authentic compounds. If they were not available commercially, they were synthesized in the laboratory.

A total of 220 compounds were identified as VDP produced during deep-fat frying (Table VI). The amount of them was mostly in ppm. However, many of them were of known toxic properties.

Among the VDP identified, the unsaturated lactones were of particular interest. The γ-lactones with unsaturation at the 2 or 3 position, viz. 4-hydroxy-2-nonenoic acid, lactone and 4-hydroxy-3-nonenoic acid, lactone, impart a characteristic deep-fat fried flavor to cottonseed oil when added at 2.5 ppm (12). The responses of the panel to the description of the cottonseed oil, plus 4-hydroxy-2-nonenoic acid, lactone, included nutty, fried fat notes, plus a butter-like note.

Organoleptic evaluation also showed that 4-hydroxy-2-nonenoic acid, lactone and 4-hydroxy-2-octenoic acid, lactone had an adverse effect upon the flavor of margarine. However, addition of 2.5 ppm of 4-hydroxy-3-nonenoic acid and 4-hydroxy-3-octenoic acid improved the flavor of margarine. The former could also improve the flavor of a snack food.

This observation explains well some previously published results. 4-Hydroxy-2-nonenoic acid, lactone was found by Krishnamurthy and Chang (7) in corn oil after it was simulatedly deep-fat fried. However, such lactones were not found by Reddy et al. (9) in hydrogenated cottonseed oil after it was simulatedly deep-fat fried under identical conditions. Since such lactones are

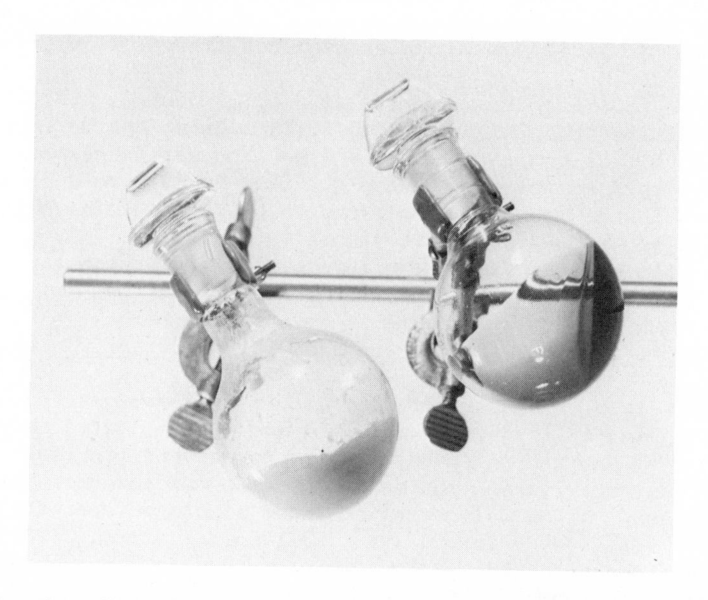

Figure 5. Thermally oxidized and polymerized materials (bottom) *isolated from a hydrogenated shortening that had been used for frying for one week*

Figure 6. Deep-fat frying of potatoes in fresh corn oil (left) *and corn oil that had been used for frying at 185°C for 72 hr*

ascertained to play an important role in deep-fat fried flavor, they might be the reason why Blumenthal, et al. (1) found that vegetable oils, such as corn oil, had a more desirable flavor than hydrogenated fat, such as hydrogenated soybean oil, after both were subjected to simulated deep-fat frying under identical conditions.

More recently, Thompson, et al. (10) isolated, fractionated, and identified the VFC produced by trilinolein, and May, et al. (11) identified those by triolein when each was subjected to simulated deep-fat frying, separately. They found more unsaturated lactones in the decomposition products of trilinolein, but less in those of triolein; both in number and in amount. This may indicate that some linoleic acid is necessary in frying fat if the characteristic deep-fat fried flavor is desired.

Nonvolatile Decomposition Products (NVDP) and Minor Constituents

The flavor chemistry of deep-fat fried flavor is, however, more complicated than the simplified theory postulated above. During deep-fat frying, NVDP are also formed and linoleate being more prone to oxidation tends to form more oxidized and polymerized esters. After being treated under simulated deep-fat frying conditions at 185°C. for 74 hrs., trilinolein formed 26.3% of non-urea-adduct-forming esters, while under the same conditions, triolein only formed 10.8% and tristearin 4.2% of such esters.

The thermally oxidized and polymerized materials, as separated by column chromatography from a shortening which had been used for deep-fat frying for a week, is shown in Figure 5. The thermally oxidized and polymerized materials were found by many researchers as having an adverse effect on human health (13). They could also cause practical problems in deep-fat frying such as foaming when the moist food is dipped into the heated oil (Figure 6).

In addition, minor constituents in the fats and oils may produce additional flavor compounds when they are used for deep-fat frying. A typical example is that french fries prepared with beef tallow shortening are well known to have a more desirable flavor than those prepared with vegetable shortenings.

LITERATURE CITED

1. Blumenthal, Michael M., J. R. Trout and S. S. Chang. Journal of the American Oil Chemists' Society (1976) 53(7) 496-501
2. Deck, Rudolph E., Jan Pokorny and S. S. Chang. Journal of Food Science (1973) 38(2) 345-349
3. Lee, Shu-Chi, B. R. Reddy and S. S. Chang. Journal of Food Science (1973) 38(5) 788-790
4. Krishnamurthy, R. G., Tsukasa Kawada and S. S. Chang. Journal of the American Oil Chemists' Society (1965) 42(10) 147-154

5. Paulose, M. M. and S. S. Chang. Journal of the American Oil Chemists' Society (1973) 50(5) 147-154
6. Kawada, Tsukasa, R. G. Krishnamurthy, B. D. Mookherjee, and S. S. Chang. Journal of the American Oil Chemists' Society (1967) 44(2) 131-135
7. Krishnamurthy, R. G. and S. S. Chang. Journal of the American Oil Chemists' Society (1967) 44(2) 136-140
8. Yasuda, Kosaku, B. R. Reddy and S. S. Chang. Journal of the American Oil Chemists' Society (1968) 45(9) 625-628
9. Reddy, B. R., K. Yasuda, R. G. Krishnamurthy, and S. S. Chang. Journal of the American Oil Chemists' Society (1968) 45(9) 629-631
10. Thompson, J. A., W. A. May, M. M. Paulose, R. J. Peterson, and S. S. Chang. Journal of the American Oil Chemists' Society (submitted for publication)
11. May, W. A., R. J. Peterson and S. S. Chang. Journal of the American Oil Chemists' Society (submitted for publication)
12. May, W. A., R. J. Peterson and S. S. Chang. Journal of Food Science (submitted for publication)
13. Artman, Neil R. Advances in Lipid Research (1969) 7 245-330

ACKNOWLEDGEMENTS

The data presented in this paper are the results of research conducted by M. M. Blumenthal, T. Kawada, R. G. Krishnamurthy, Shu-Chi Lee, W. A. May, Braja Mookherjee, M. M. Paulose, B. R. Reddy, J. A. Thompson, and K. Yasuda.

RECEIVED December 22, 1977

3

Volatiles from Frying Fats: A Comparative Study

W. W. NAWAR, S. J. BRADLEY, S. S. LOMANNO,
G. G. RICHARDSON, and R. C. WHITEMAN

Department of Food Science and Nutrition, University of Massachusetts,
Amherst, MA 01003

Since the early 1940's when Farmer and his co-workers advanced their theory of fat oxidation, significant achievements have been made in instrumental developments and analytical techniques. Consequently, large numbers of oxidative decomposition products from heated or oxidized lipids have been identified, and the number of such compounds continues to increase every day, as our instrumental capabilities continue to become more and more sensitive and sophisticated.

Unfortunately, however, flavor research has not kept pace with chemical identification. Our knowledge regarding the exact role of these many compounds, in relation to specific flavors or off-flavors, is far from being adequate. A thorough examination of the reported literature reveals much ambiguity and raises many questions. We know very little about the exact difference between an unpleasant rancid flavor and a pleasant fried odor - from the standpoint of their specific chemical make-up. What exactly is the chemical difference between several used oils, all good, but having different flavor characteristics and used oils, all bad, but having different objectionable odors?

We do know that the major oxidative decomposition products in natural fats are mostly those resulting from breakdown of linoleates. These are essentially the saturated aldehydes, the alkenals and the dienals. But, the same major compounds are present above their threshold levels in rancid beef, boiled chicken, frozen pork, good used frying oil, bad overused shortenings, etc. In fact, many of the same compounds are also present but in much smaller amounts in fresh unused fats. Is the difference in amounts of the same compounds responsible for flavor differences between the oils? Does a given set of oxidation products produce unpleasant rancid flavors when present in certain concentrations but give rise to pleasant flavors if present in different concentrations? If so, what concentrations are critical for which flavors? Are there specific key components responsible for each flavor or each variation of the same flavor? If so, what are these key compounds? Why is it that a compound can be

0-8412-0418-7/78/47-075-042$05.00/0

blamed for an off-flavor by one investigator, but given credit for producing a pleasant flavor by another author?
What is the meaning of statistical correlations between specific compounds and specific flavors, in terms of the causative agents? For example, positive correlation between pentane and rancidity is an important fact, but pentane does not produce rancid flavor.

What is the role of the less volatile oxidative compounds which we know are present but are difficult to identify and even more difficult to quantitate?

What is the role of the non-oxidative, strictly thermal, decomposition products, the presence of which must be superimposed on that of the oxidation compounds? How do all of these many compounds interact with each other, qualitatively and quantitatively, to effect certain flavor responses?

Why is there so much contradiction in the literature, particularly when it comes to the effects of operating parameters on the decomposition of fats? The introduction of moisture, for example, during the frying process was reported by some workers, to markedly accelerate the deterioration of shortenings and increase the quantity of carbonyl compounds produced, while other workers report the exact opposite with moisture exhibiting a protective effect.

How reliable is our identification of new compounds at, or below, the ppb levels and how accurate are our quantitative measurements? As the analytical techniques become more sensitive, mistakes can be made more easily. For example, interference by misleading artifacts or contaminants becomes more likely and quantitative analysis becomes less accurate. It is not uncommon to see gas chromatographic (GC) quantitative data calculated without use of internal standards, recovery determinations, proper controls or appropriate correction factors. Component peak overlap in GC analysis has been, and continues to be, a serious threat to both qualitative and quantitative analysis. In many instances identification of certain decomposition products reported in heated fats could not be repeated in our laboratory when milder techniques of isolation were used. In other cases, identification of major decomposition products using GC packed columns were proved erroneous when capillary columns were employed.

Recently we began a study to investigate the relationship between various operating parameters and chemical changes in frying fats. In this report quantitative data are provided to demonstrate the effects of such factors as type of oil, length of use, temperature of frying and introduction of moisture on the major volatile compounds produced. Of course the total decomposition pattern in frying is not limited to the volatile profiles examined in the present paper. The importance of the non-volatile and the minor breakdown products to the quality of the frying oil must be recognized.

Experimental

Samples of oil were placed in round bottom flasks and heated in air with the aid of an oil bath. During heat treatment the oil samples were continually stirred with a magnetic stirrer. For introduction of moisture a burette and a microsyringe were used as shown in Figure 1. The level of water in the burette was adjusted so as to provide a flow of 5 microliters of water per ml oil per hour. After heat treatment the volatile decomposition products were collected for qualitative and quantitative analysis.

The following guidelines were strictly followed to insure the absence of artifacts and reliability of quantitative data:

1. Relatively small samples of oils (1-10g) were used.
2. Only simple conditions were employed for the collection of volatiles, i.e., the short path high-vacuum distillation described previously (1).
3. To minimize GC peak overlap the volatiles were prefractionated on silica into two fractions, one polar and the other non-polar, and each fraction was separated on a 500 ft. .02 in. Carbowax capillary column.
4. In addition to GC retention times and mass spectrometric analysis for all samples, at least one microchemical technique (e.g. hydrogenation, reaction with 2,4-dinitrophenyl hydrazine, etc.) was used to confirm identification.
5. Quantitative measurements were conducted in quadruplicates with the aid of two internal standards, one polar and one non-polar. These were placed in the oil after the heat treatment and, thus, allowed to suffer the same fate as the other volatiles throughout the procedures of collection, pre-fractionation and GC separation.
6. A correction factor for each compound was used to correct for differences in analytical behavior.
7. Identical conditions were used throughout the study with a weekly check on the performance of the capillary column, GC response and the mass spectrometer using a standard quantitative mixture of hydrocarbons.

Results and Discussion

Gas chromatographic patterns of the major non-polar and polar volatile compounds produced in corn, soybean and coconut oils by heating at 185 C for 2 hrs. are shown in Figures 2 and 3. Quantitative data (averages of quadruplicate separate heating experiments) for corn, soybean and coconut oils heated at 185 C and 250 C are given in Table I and II. In a series of publications, Chang and co-workers reported the identification of a much larger number of volatile decomposition products in

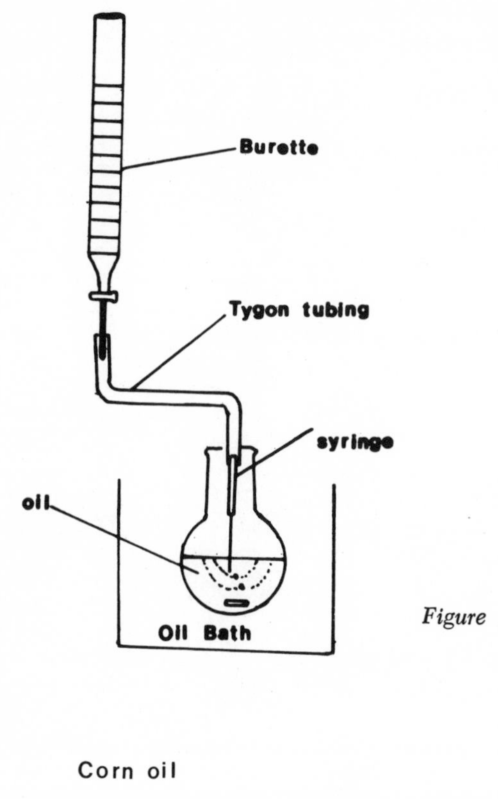

Figure 1. Apparatus for introduction of moisture

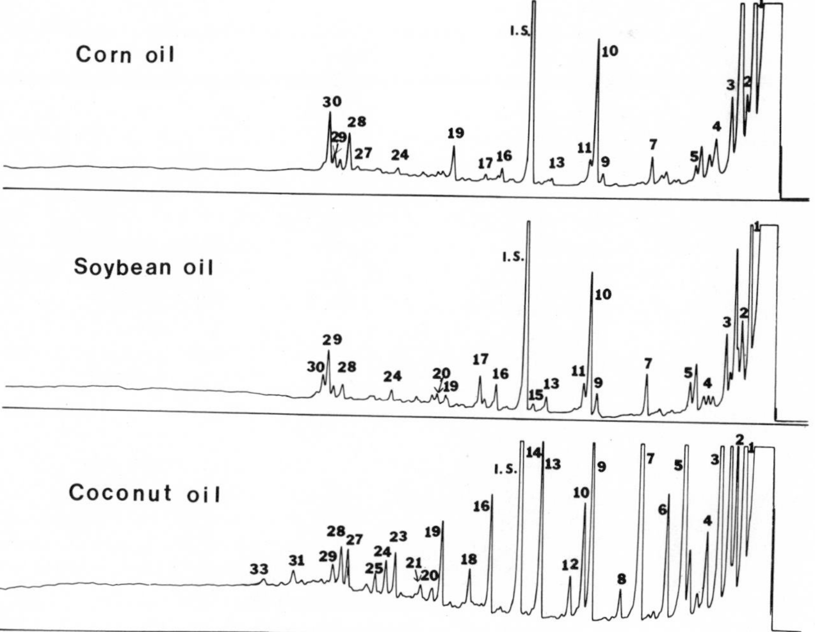

Figure 2. GC analysis of the nonpolar volatile components in corn, soybean, and coconut oils heated for 2 hr at 185°C

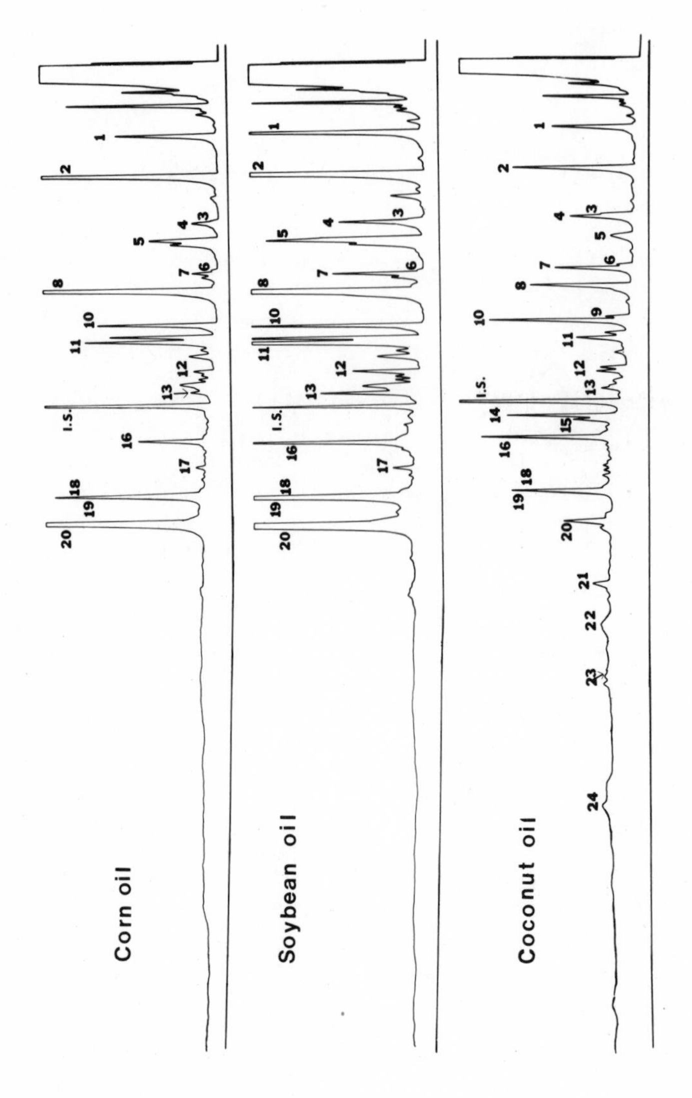

Figure 3. GC analysis of polar volatile components in corn, soybean, and coconut oils heated for 2 hr at 185°C

Table I. Quantitative Analysis (mg volatile/Kg oil) of the major Non-Polar Compounds in Three Vegetable Oils After Heating for 2 hrs. at 185 C and 250 C.

Peak No. in Fig. 2	Compound	Corn		Soybean		Coconut	
		185 C	250 C	185 C	250 C	185 C	250 C
1	octane	36.	48.	23.	31.	68.	100
2	1-octene	0.7	3.	0.9	3.2	1.6	4.0
3	nonane	1.1	2.2	1.9	4.2	14.	39.
4	1-nonene	0.5	1.8	0.4	1.5	1.3	2.5
5	decane	0.9	1.5	1.0	2.1	12.	26.
6	1-decene	n.d.[a]	n.d.	n.d.	n.d.	2.6	10.
7	undecane	1.8	2.9	1.3	2.3	13.	58.
8	1-undecene	n.d.	n.d.	n.d.	n.d.	0.6	2.0
9	dodecane	0.5	1.8	0.8	1.8	5.3	9.9
10	pentylfuran	9.5	11.	3.2	2.9	1.6	1.0
11	Int. dodecene[b]	2.5	7.1	1.2	3.2	n.d.	n.d.
12	1-dodecene	n.d.	n.d.	0.4	0.5	1.0	3.4
13	tridecane	0.4	0.7	0.6	1.0	n.d.	n.d.
14	butyl benzene	n.d.	10.	n.d.	3.3	n.d.	13.
15	Int. tridecene	n.d.	2.5	n.d.	2.1	n.d.	n.d.
16	tetradecane	0.7	1.4	1.0	1.2	2.6	4.1
17	Int. tetradecene	0.2	1.3	0.2	1.2	0.1	0.4
18	1-tetradecene	0.1	0.5	n.d.	0.3	0.8	2.0
19	pentadecane	0.7	1.0	0.5	0.7	1.7	4.6
20	Int. pentadecene	0.3	10.	0.4	6.1	0.2	1.6
21	1-pentadecene	0.2	0.7	0.3	0.3	0.2	n.d.
22	pentadecadiene	0.2	12.	n.d.	6.6	n.d.	n.d.
23	hexadecane	n.d.	0.7	0.3	0.3	0.9	1.2
24	Int. hexadecene	0.4	1.8	0.5	1.6	0.8	1.8
25	1-hexadecene	0.1	7.0	0.2	2.3	0.4	0.8
26	hexadecadiene	n.d.	4.7	n.d.	2.1	n.d.	n.d.
27	heptadecane	0.2	n.d.	0.3	0.2	0.6	1.7
28	Int. heptadecene	0.9	1.8	1.1	1.4	0.9	3.8
29	1-heptadecene[c]	1.2	8.6	1.4	8.5	n.d.	0.9
30	heptadecadiene	1.2	2.7	0.9	1.3	0.3	0.4
31	octadecane	n.d.	n.d.	0.1	0.1	0.2	0.2

a. Not detectable under the experimental conditions used.
b. Internally unsaturated: double bond not in terminal position.
c. This component overlapped with another unknown compound having a molecular ion at m/e 266.

Table II. Quantitative Analysis (mg volatile/Kg oil) of the Major Polar Volatile Compounds in Three Vegetable Oils After Heating for 2 hrs. at 185 C and 250 C.

Peak No. in Fig. 3	Compound	Corn		Soybean		Coconut	
		185 C	250 C	185 C	250 C	185 C	250 C
1	pentanal	44.	37.	39.	47.	31.	29.
2	hexanal	120	120	94.	100	42.	43.
3	2-heptanone	3.6	7.5	3.2	2.6	11.	14.
4	heptanal	12.	24.	12.	28.	23.	42.
5	hexenal pentylfuran [b]	42.	45.	44.	42.	17.	7.0
6	2-octanone	1.9	3.7	1.5	2.9	5.9	5.8
7	octanal	8.1	10.	12.	16.	18.	31.
8	heptenal	130	140	140	130	37.	28.
9	2-nonanone	n.d.[a]	n.d.	n.d.	n.d.	7.4	12.
10	nonanal	49.	50.	47.	53.	52.	59.
11	octenal	54.	43.	64.	50.	18.	13.
12	decanal	6.3	5.9	8.3	7.2	9.4	23.
13	nonenal	15.	25.	31.	34.	12.	10.
14	2-undecanone	n.d.	n.d.	n.d.	3.5	31.	41.
15	undecanal	n.d.	n.d.	n.d.	3.9	12.	20.
16	decenal	29.	33.	42.	40.	43.	27.
17	dodecanal	n.d.	n.d.	n.d.	2.3	2.4	10.
18	undecenal	24.	26.	30.	28.	39.	21.
19	decadienal[c]	84.	21.	100	29.	n.d.	n.d.
20	decadienal[d]	250	72.	290	100	97.	32.
21	γ-octalactone	n.d.	n.d.	n.d.	n.d.	7.6	6.2
22	δ-octalactone	n.d.	n.d.	n.d.	n.d.	2.7	1.5
23	γ-nonalactone	n.d.	n.d.	n.d.	n.d.	3.7	4.2
24	γ-decalactone	n.d.	n.d.	n.d.	n.d.	5.6	5.5
25	δ-decalactone	n.d.	n.d.	n.d.	n.d.	1.8	1.5
26	γ-undecalactone	n.d.	n.d.	n.d.	n.d.	2.9	2.3
27	γ-dodecalactone	n.d.	n.d.	n.d.	n.d.	33.	35.

a. not detectable under the experimental conditions used.
b. peaks not well resolved.
c. tr,cis-2,4-decadienal.
d. tr,tr-2,4-decadienal.

frying fats (2-6). It should be pointed out, however, that these workers were obliged to pool the volatiles from several frying treatments, collections and fractions, with the amount of oil used totaling approximately 31 kilograms. In the present work, all the volatiles were collected, separated, identified and measured during one analysis involving a sample of only 1-10 grams. This was important to insure reliable quantitative determinations of each of the volatiles and to minimize artifacts.

In general, the non-polar compounds are present in heated fats in much lower quantities than the polar components. Typically the non-polar products produced at 185 C consist of a homologous series of the n-alkanes from C_8 to C_{18}, a series of 1-alkenes from C_8 to C_{17}, the longer chain alkadienes from C_{15} to C_{17} and pentylfuran. The hydrocarbons shorter than C_8 were not determined in this work. Under the conditions used here for pre-fractionation, pentylfuran partitions between the polar and non-polar fractions. Octane and pentylfuran represent the major non-polar compounds in corn and soybean oils. Heated coconut oil, on the other hand, contains higher amounts of the shorter chain hydrocarbons, a phenomenon reflecting its unique fatty acid composition (i.e. higher concentration of the shorter-chain saturated fatty acids). Most of the non-polar compounds are produced in greater amounts when the fat is heated at higher temperatures. In addition, when heated at 250 C all three fats contained some new compounds which were absent or barely detectable after heating at 185 C. The most important of these compounds is butyl benzene (Table I).

The pattern of the major polar compounds in heated fats is also typical and consists of a series of alkanals, alkenals and dienals as well as smaller amounts of some methyl ketones. In corn and soybean oil, hexanal, heptanal and tr,tr-2,4-decadienal are the three aldehydes produced in the greatest quantity. Hexanal and 2,4-decadienal are the two major aldehydes expected from decomposition of the conjugated 9- and 13-hydroperoxides, the initial products of linoleate autoxidation, while octanal, nonanal, 2-decenal and 2-undecenal are the major aldehydes predicted from oleates (7). It can be seen that aldehydes other than those expected on the basis of theory are produced, in some cases at relatively large quantities. Thus, the fact that heptenal represents such a major decomposition product in corn and soybean oils is difficult to explain. A similar semi-quantitative pattern of aldehydes in autoxidized sunflower oil was reported by Swoboda and Lea (8).

Although the same aldehydes are present in coconut oil, the unique fatty acid composition of this fat is again reflected in the following quantitative and qualitative aspects when compared with the other two oils. For example, in the case of coconut oil, heating produced lesser quantities of the decadienals, higher quantities of the saturated aldehydes (with the exception of hexanal and pentanal) and lower quantities of the unsaturated

Table III. Concentration of Volatile Compounds (mg/Kg) in Corn Oil after Heating at 185 C for Various Periods of Time in the Absence and Presence of Moisture.

Compound	Time of Heating					
	1 Hr.		5 Hr.		48 Hr.	
	No H_2O	H_2O	No H_2O	H_2O	No H_2O	H_2O
Nonpolar Fr.						
octane	1.8	0.5	6.1	0.6	10.	3.1
1-octene	1.7	n.d.[a]	0.8	n.d.	4.7	n.d.
nonane	0.8	0.02	0.4	0.03	3.4	0.5
decane	0.2	0.1	0.8	0.1	1.9	0.4
undecane	0.2	0.1	0.8	0.1	3.6	0.4
1-undecene	0.1	0.1	0.3	0.1	0.3	0.04
dodecane	0.1	0.1	0.2	0.1	2.4	0.3
pentylfuran	1.9	0.4	7.9	0.7	90.9	6.3
int. dodecene	0.5	0.1	1.0	0.2	10.9	0.5
1-dodecene	0.1	n.d.	0.1	n.d.	1.1	n.d.
tridecane	n.d.	0.03	n.d.	0.04	1.4	0.2
1-tridecene	n.d.	n.d.	0.1	n.d.	0.7	n.d.
tetradecane	0.1	0.1	0.3	0.1	1.6	0.3
int. tetradecene	n.d.	n.d.	0.1	n.d.	2.3	n.d.
1-tetradecene	0.1	0.02	0.2	0.03	1.3	0.03
pentadecane	0.3	0.1	0.7	0.1	1.0	0.3
int. pentadecene	0.1	n.d.	0.4	n.d.	6.1	n.d.
1-pentadecene	0.4	0.1	0.4	0.2	0.7	0.3
pentadecadiene	0.2	0.1	0.5	0.2	3.3	0.3
hexadecane	0.2	0.02	0.1	0.03	0.2	0.1
int. hexadecene	0.2	n.d.	0.2	0.1	2.2	0.4
1-hexadecene	n.d.	0.02	0.1	0.1	1.1	0.2
hexadecadiene	n.d.	n.d.	0.1	0.04	1.0	0.1
heptadecane	0.2	0.1	0.2	0.1	0.4	0.3
int. heptadecene	0.4	0.1	1.0	0.2	1.4	0.7
1-heptadecene	0.2	0.1	1.0	0.6	2.4	0.6
heptadecadiene	0.6	0.2	1.8	0.4	2.3	0.7
octadecane	0.3	0.1	0.1	0.1	0.2	0.2
Polar Fr.						
pentanal	2.3	2.3	7.3	b	66.	6.5
hexanal	13.	7.0	42.	b	310	42.
heptanal	1.6	0.9	6.6	b	68.	6.2
hexenal pentylfuran	8.1	2.7	17.7	b	130	13.5
octanal	1.0	0.4	2.4	b	29.	2.6
heptenal	10.	10.	47.	b	360	16.
nonanal	5.2	3.1	12.	b	100	17.
octenal	4.3	4.0	20.	b	180	14.
decanal	0.9	0.5	2.2	b	20.	2.4
nonenal	3.0	1.2	16.	b	120	19.

Table III. Contd.

	Time of Heating					
	1 Hr.		5 Hr.		48 Hr.	
	No H_2O	H_2O	No H_2O	H_2O	No H_2O	H_2O
undecanal	0.4	0.2	0.8	b	8.6	1.6
decenal	2.7	3.2	8.9	b	100	23.
dodecanal	n.d.	0.1	0.4	b	4.7	0.8
undecenal decadienal	7.0	13.	20.	b	210	52.
decadienal	32.	72.	100	b	640	190

a. not detectable under the experimental conditions used.
b. polar compounds were not measured at the 5 hr. interval.

aldehydes. In addition, the formation of a number of methyl ketones and gamma and deltalactones is typical for coconut oil.

The most remarkable effect of heating temperature on the polar compounds produced is the significant reduction in the amounts of the decadienals present when the fats were heated at 250 C as compared to heating at 185 C. It is possible that these compounds undergo further decomposition at the higher temperatures.

Most of the volatile compounds discussed above were present in unheated control oils at concentrations lower than 0.1mg volatile/Kg in case of the non-polar fraction and 0.2mg/Kg for the polar compounds.

The quantitative effects of heating time and introduction of moisture on the major polar volatile components are given in Table III. In this experiment 100g samples were heated and 10ml aliquots removed periodically for quantitative analysis. The observation that lower values were obtained in this experiment after 5 hrs. of heating as compared to the amounts reported in Tables I and II for 1 hr. heatings can be attributed to a faster oxidation expected in the small samples which have larger surface to volume ratios. This effect was evident in the 1 hr. heatings where 10g per flask were used. In general the individual volatile components increased with continued heating at 185 C up to 48 hrs. However, a drastic reduction in the quantities of these compounds was observed, particularly in case of the polar compounds (Table III), when moisture was introduced as compared with dry heating, thus indicating a definite protective effect of moisture. These results support the observations of Peled and co-workers (9) but are in contradiction with other studies in which water was reported to accelerate thermal oxidative deterioration of oils (10,11). These workers, however, did not measure individual volatile compounds. They evaluated their oils through such tests as acid, hydroxyl, TBA, extinctions at 232 and 460 nm, amounts of polymers, and carbonyl values. The protective effect observed in the present study is probably due to either the stripping of volatiles by the generated steam, the displacement of atmospheric oxygen by the inert steam or a combination of both.

To compare the patterns obtained in the laboratory heating experiments described above with those arising from actual frying operations, samples were periodically taken from a frying vat in the University Dining Kitchens and analyzed. The same vat, with a shortening capacity of 50 gallons, was used only 3 days each week for the frying of potatoes, chicken, veal patties and shrimp. The shortening was filtered daily, and replenished with approximately 5 lb. shortening whenever appropriate, but otherwise was not changed throughout the experiment. Table IV shows that the qualitative pattern of the major volatile decomposition products present in the used shortening were essentially the same as that obtained from heating corn or soybean oils. Quantitatively, however, it can be seen that even after 12 weeks of use, the amounts

Table IV. Concentration of Volatile Compounds (mg/Kg) in University Kitchen Shortening After Successive Periods of Use.

Compound	Time In Use (weeks)			
	6	8	10	12
Nonpolar Fr.				
nonane	.45	.65	n.d.[a]	.83
1-nonene	.12	n.d.	.11	.17
decane	.29	.39	.10	.32
undecane	.42	.43	.11	.31
dodecane	.37	.34	.12	.26
pentylfuran	.43	.19	.40	.43
1-tridecene	.07	n.d.	.03	.06
tetradecane	.81	.41	.12	.50
1-tetradecene	.20	.11	n.d.	.11
pentadecane	2.2	.65	.42	1.4
int. pentadecene	.93	.53	.21	.5
1-pentadecene	.29	n.d.	.44	.18
pentadecadiene	.2	n.d.	.11	.12
hexadecane	.51	.3	.24	.41
int. hexadecene	1.5	.60	.35	2.3
1-hexadecene	.17	n.d.	.16	.13
heptadecane	1.6	.71	.61	1.4
int. heptadecene	4.2	1.1	1.9	3.1
1-heptadecene	.96	.57	.57	1.6
heptadecadiene	.44	.15	.51	.63
octadecane	.34	.21	.14	.53
Polar Fr.				
hexanal	1.1	.61	.3	1.7
2-heptanone heptanal	.43	.28	.11	.57
hexenal pentylfuran	.36	.43	.16	.8
2-octanone octanal	.54	.41	.10	.74
heptenal	1.8	1.4	.31	2.0
nonanal	3.	2.	.55	3.9
octenal	1.5	1.2	.35	2.2
decanal	.23	.1	.12	.28
nonenal	2.7	2.	.59	2.2
decenal	3.5	3.	.74	4.6
undecenal decadienal	3.7	3.	.78	5.5
decadienal	11.	5.3	3.	20.

a. not detectable.

of the volatile decomposition products were lower than those found in our laboratory controlled experiments in which the oils were heated for much shorter times. No change in qualitative pattern or significant accumulation of any of the volatile components can be related to time of use. Obviously, regular dilution with fresh oil to replenish the shortening present in the frying vat, as well as the steam generated by the food fried in the shortening, are responsible for the low amounts of these volatiles.

Samples were also taken at random from various commercial frying operations and the volatiles quantitatively analyzed. In spite of the fact that the shortenings used, the foods fried and the age of the oils varied widely, the qualitative pattern of the decomposition products was again typical of that obtained from heated corn oil. In addition, various other components including nitrogen-containing compounds were detected. These were undoubtedly contributed by the foods fried in the oils. Quantitatively, the decomposition products were almost always present in lower quantities than those found in corn oil continuously heated, for a very short time. This is illustrated in Table V where the amounts of some volatiles in shortenings obtained from typical frying operations are compared to the volatiles produced in corn oil heated without water for only one hr. or with water for 70 hrs. The data shows clearly that continous heating of oils produces far greater quantities of these volatiles than normal frying conditions, and that the amounts of volatiles studied here cannot be used as an indicator of the extent to which a commercial frying oil had been used.

Work is in progress in our laboratory to investigate the effects of other frying parameters on the chemical changes in heated oils.

Table V. Concentration of Some Volatiles in Used Frying Shortenings as Compared to Those in Heated Corn Oil (mg/Kg).

	Corn 1 Hr.	Corn/H_2O 70 Hr.	Univ. Kitchens 12 Wks.	Chicken Frying 10 Days
Hexanal	13.	11.	1.7	2.2
Heptenal	10.	4.2	2.1	2.2
Octenal	4.3	4.9	2.2	2.7
Decadienal	6.5	7.4	5.0	5.8
Decadienal	32.	20.	20.	12.
Octane	1.8	1.2		4.4
Undecane	.24	.12	.48	.6
Pentylfuran	1.9	3.6	.4	.9
Pentadecane	.3	.2	1.4	1.3

Literature Cited

1. Nawar, W. W., Champagne, J. R., Dubravcic, M. F., Letellier, P. R. J. Agr. Food Chem. (1969) 17, 645.
2. Krishnamurthy, R. G., Kawada, T. and Chang, S. S. J. Am. Oil Chemists' Soc. (1965) 42, 878.
3. Kawada, T., Krishnamurthy, R. G., Mookherjee, B. D. and Chang, S. S. J. Am. Oil Chemists' Soc. (1967) 44, 131.
4. Krishnamurthy, R. G. and Chang, S. S. J. Am. Oil Chemists' Soc. (1967) 44, 136.
5. Yasuda, K., Reddy, B. and Chang, S. S. J. Am. Oil Chemists' Soc. (1968) 45, 625.
6. Reddy, B. R., Yasuda, K. and Chang, S. S. J. Am. Oil Chemists' Soc. (1968) 45, 629.
7. Schultz, H. W., Day, E. A. and Sinnhaber, R. O. "Symposium on Foods. Lipids and Their Oxidation". A.V.I. Pub. Co., Connecticut, 1962.
8. Swoboda, P. A. T. and Lea, C. H. J. Sci. Fd. Agric. (1965) 16, 680.
9. Peled, M., Gutfinger, T. and Letan, A. J. Sci. Fd. Agric. (1975) 26, 1655.
10. Dornseifer, T. P., Kim, S. C., Keith, E. S., Powers, J. J. J. Am. Oil Chemists' Soc. (1965) 42, 1073.
11. Perkins, E. G., VanAkkern, L. A. J. Am. Oil Chemists' Soc. (1965) 42, 782.

Acknowledgments

Paper No. 2190, Massachusetts Agricultural Experiment Station, University of Massachusetts. This work was supported in part from University of Massachusetts Expt. St. Project No. 198. The authors are grateful to Henry Wisneski, Susan Kakley and Susan Henderson for their assistance.

RECEIVED December 22, 1977

4

Generation of Aroma Compounds by Photo Oxidation of Unsaturated Fatty Esters

BRAJA D. MOOKHERJEE and ROBERT W. TRENKLE

International Flavors and Fragrances, Inc., 1515 Highway #36,
Union Beach, NJ 07735

Of all essential oils, jasmine is probably the most precious and most widely used in fine quality perfumes. The fragrance of jasmine is a secret which still eludes the skill of chemist and perfumer who would reproduce it synthetically.

The pretty jasmine flower originates in the lower valleys of the Himalayas of northern India. The Moors brought this plant to Spain, and from there it started to spread in the sixteenth century along the Mediterranean coast. All jasmine extracts are obtained from the flowers of Spanish jasmine (*Jasminum grandiflorum*, L.). The yield of absolute from the flower is very low. One kilogram, (i.e., about 8,000 flowers) yields only 1.5 g of absolute. The price of jasmine absolute ranges up to more than $2000 per pound, depending on its source.

Due to its importance in the fragrance industry, jasmine oil has been investigated by several research teams from 1899 to 1973 and some forty components are recorded in the literature [1,2]. These include eight carbonyls, viz., cis-jasmone I,[3] benzaldehyde[4], methyl jasmonate[5], vanillin[6], methyl heptenone[5], 6, 10, 14-trimethyl pentadecanone-2[6], jasmin-keto-lactone[7], and N-acetyl methyl anthranilate[8], Even though all of these components contribute to the total olfactive impression, only cis-jasmone I and methyl jasmonate II have the typical jasmine odors.

In 1967 we undertook the task of analyzing jasmine absolute in the hope that we would be able to discover new and interesting components which would help to create a superior synthetic jasmine.

Within a year, through careful analytical procedure we identified a novel chemical, dehydro methyl jasmonate III possessing an interesting jasmine odor.

In 1964, Dr. Demole[9], the discoverer of methyl jasmonate II, prposed that methyl jasmonate is formed

0-8412-0418-7/78/47-075-056$05.00/0

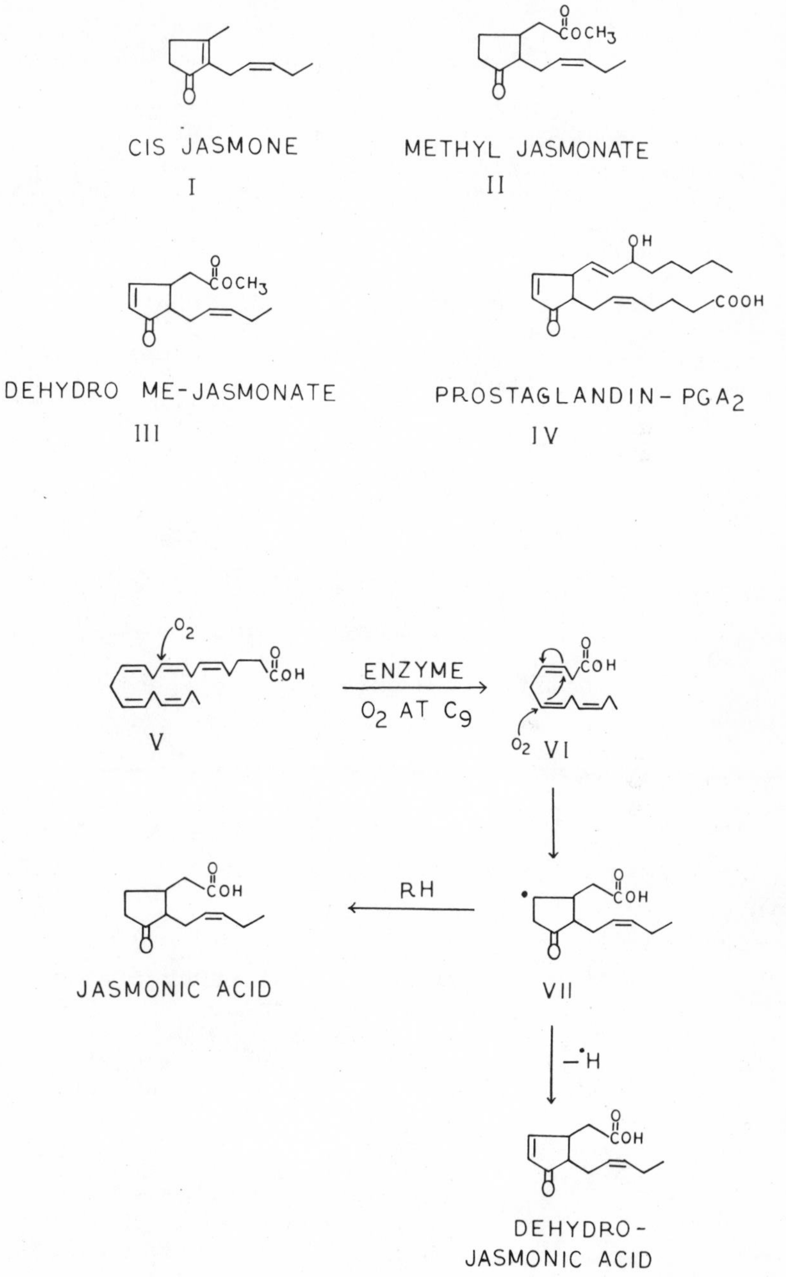

CIS JASMONE
I

METHYL JASMONATE
II

DEHYDRO ME-JASMONATE
III

PROSTAGLANDIN-PGA_2
IV

V

ENZYME
O_2 AT C_9

VI

RH

JASMONIC ACID

VII

$-\dot{H}$

DEHYDRO-
JASMONIC ACID

Figure 1.

by the condensation of two poly beta-diketone chains. But the striking structural similarity of our dehydro methyl jasmonate III to prostaglandin-PGA_2 IV strongly suggests that the jasmonate compounds in jasmine oil are probably formed by the oxidative (enzyme) cyclization of polyunsaturated fatty acids (Fig. 1).

Enzymatic (lipoxidase) oxidation of all cis-5,8,11, 14,17-eicosa pentanoic acid VI, which could undergo oxidative cyclization, as in the case of prostaglandin[10], to form a cyclopentanone free radical VII. This radical could not only abstract a proton to form jasmonic acid but also could undergo disproportionation to form dehydro jasmonic acid.

During the analysis of jasmin oil we detected the presence of many polyunsaturated fatty acids in addition to linolenic acid. Therefore, in order to test the above hypothesis, we choose commercially available polyunsaturated mixed fatty acids containing 50% linolenic acid.

The mixed fatty acids known as vegacid was obtained from Archer Danial Midland Co. This material was esterified with methanol in the presence of sulphuric acid. The methyl esters of these mixed acids were photo oxidized by bubbling dry oxygen (15 ml/minute) for 130 hours under 450 watt high pressure Hanovia UV lamp. The composition of this material is as follows:

	Starting Esters %	Oxidized Esters %
methyl plamitate	6.6	14.6
methyl oleate	23.0	42.4
methyl lenoleate	19.7	14.4
methyl linolenate	50.2	14.5

The oxidized esters oil were then subjected to series of careful distillation, column chromatography and gas liquid chromatographic procedures. By this method we isolated a peak whose IR, NMR and MS are superimposable with those of the authentic methyl jasmonate. The odor properties of this isolated material is also the same as that of the synthetic methyl jasmonate.

Conclusion.

By this way we proved for the first time that an odoriferous cyclopentanone molecule could be generated by the photo-oxidative degradation of unsaturated fatty acids.

LITERATURE CITED

1. Van der Gen., Parf. Cosm. Sav. France, (1973), 2, 356.
2. Polak, E. H., Cosmetics and Perfumery, (1973), 88, 46.
3. Hesse, A., Ber., (1899), 32, 2611.
4. Naves, Y. R. and Grampoloff, A. V., Helv. Chem. Acta., (1942), 25, 1500.
5. Demole, E., Lederer, E. and Mercier, D., Helv. Chem. Acta, (1962), 45, 675.
6. Demole, E., Helv. Chem. Acta, (1962) 45, 1951.
7. Naves, Y. R., Grampoloff, A. V., and Demole, E., Helv. Chem. Acta (1963) 46, 1006.
8. Demole, E., "Thin Layer Chrom., Proc. Symp., Rome. (1963, published 1964), 45.
9. Demole, E., Willhalm, B. and Stole M., Helv. Chem. Acta (1964) 47, 1152.
10. Burgstrom, S. and Samuelsson, B., Endeavour (1968), 27, 109.

RECEIVED December 22, 1977

5

Instrumental Analysis of Volatiles in Food Products

HAROLD P. DUPUY, MONA L. BROWN, MICHAEL G. LEGENDRE, JAMES I. WADSWORTH, and ERIC T. RAYNER

Southern Regional Research Center, Agricultural Research Service, U. S. Department of Agriculture, New Orleans, LA 70179

In recent years, considerable effort has been expended to examine the volatiles and trace components that characterize and contribute to food flavors. Some early attempts to measure food volatile components by gas chromatographic methods consisted of analyzing headspace vapors to detect vegetable and fruit aromas (1) and volatiles of various food products (2). These methods, however, require special preparation of the sample and subsequent transfer of a vapor aliquot to the gas chromatograph. Extraction and distillation techniques have been proposed to provide quantities of volatiles sufficient for instrumental detection and analysis (3,4,5,6). These methods are complex, tedious, time-consuming operations that may also produce artifacts. More recently, a direct gas chromatographic procedure was reported for the examination of volatiles in salad oils and peanut butters (7,8). The method does not require prior enrichment of volatiles and is rapid and efficient. Analysis of flavor-scored soybean oils by this direct chromatographic method, or with suitable modifications (9,10,11), has shown that their flavor quality can be measured by instrumental means. Our work provides additional evidence of the applicability of this rapid, unconventional gas chromatographic technique for analyzing flavor quality of vegetable oils, and projects its utility for other raw and processed food products.

Materials and Methods

Materials. Tenax GC[1/], 60-80 mesh (a thermostable polymer, 2,6-diphenyl-p-phenylene oxide), and Poly MPE (poly-m-phenoxylene) were obtained from Applied Science Laboratories, State

1/ Names of companies or commercial products are given solely for the purpose of providing specific informatipn; their mention does not imply recommendation or endorsement by the U. S. Department of Agriculture over others not mentioned.

0-8412-0418-7/78/47-075-060$05.00/0

College, Pa. Teflon O-rings were purchased from Alltek Associates, Arlington Heights, Ill.; sandwich-type silicone septums from Hamilton Company, Reno, Nev., and Pyrex glass wool from Corning Glass Works, Corning, N. Y. (The O-rings, septums, and glass wool were conditioned at 200 C for 16 hr prior to use.) Inlet liners, 10 X 84 mm, were cut from borosilicate glass tubing. The soybean oil samples, provided by the AOCS Flavor and Nomenclature Committee (AOCS-FNC), were experimental oils specially treated to provide a wide range of flavor variance. The oils were flavor scored by 12 taste panels from industry, academia, and government laboratories on a 1 to 10 scale. The number of sensory judges in an individual panel varied from 3 to 24, and in all, totaled 156 panelists.

Gas Chromatography (GC). A Tracor-MT-220 gas chromatograph with dual independent hydrogen flame detectors was used in conjunction with a Westronics MT22 recorder and a Hewlett-Packard Integrator, Model 3380 A. The columns were stainless steel U-tubes, 1/8 in. OD, 10 ft long, packed with Tenax GC that had been coated with 8% Poly MPE. Operating conditions were as follows: Nitrogen carrier gas, 60 ml/min in each column; hydrogen, 60 ml/min to each flame; air, 1.2 ft^3/hr (fuel and scavenger gas for both flames). Inlet temperature was 170 C. Detector was at 250 C. Column oven was maintained at 30 C during the initial 40 min hold period. After removal of the inlet liner, the column was heated to 100 C within 5 min, then programmed 3 C/min for 30 min. The final hold was at 190 C for 30 min until the column was clear. A Teflon O-ring was positioned at the bottom of the inlet of the GC to provide a leak proof seal. Electrometer attenuation was 10 X 4.

Mass Spectrometry (MS). A Hewlett Packard (quadrapole) mass spectrometer, Model 5930 A, was interfaced with a Tracor Model 222 GC. The ionization potential was 70 eV, and the scan range was from 21 to 350 amu. Scanning and data processing were accomplished with an INCOS 2000 mass spectrometer data system.

Sample Preparation and Analysis for GC. A 3-3/8 in. length of 3/8 in. OD borosilicate glass tubing was packed with volatile-free glass wool, loose enough to permit diffusion of oil throughout the packing, yet tight enough to prevent seepage of the sample from the liner onto the GC column. Clearance of 1/4 in. was allowed at the bottom of the liner and 1/2 in. at the top. The septum nut, septum, and retainer nut of the GC were removed, and the liner containing the sample was inserted in the inlet of the GC on top of the Teflon O-ring. When the retainer nut was tightened above the upper rim of the liner, a seal was formed between the base of the inlet and the lower rim. On closing the inlet system with the septum and septum nut, the carrier gas was forced to flow upward and down through the sample. This assembly has been described previously (9).

Volatiles were rapidly eluted from the sample as the carrier gas swept through the heated liner and were adsorbed on the top portion of the column, which was maintained at 30 C during the initial hold period of 40 min. The liner containing the spent sample was removed from the inlet, the integrator and programmer were turned on immediately, and the temperature was raised to 100 C in 5 min. Temperature programming was then begun. When complete, the temperature was maintained on final hold to elute and resolve the volatiles adsorbed on the column. The oven was then cooled to 30 C in preparation for the next sample.

Sample Preparation and Analysis for MS. A silicone membrane separator was used to interface the gas chromatograph with the mass spectrometer. In a typical analysis, conditions were the same as those described for GC, except that helium was the carrier gas. Volatiles that are resolved by GC temperature programming permeate the membrane and enter the mass spectrometer, where the specific peaks are identified.

Results and Discussion

The four experimental soybean oils that had been specially treated and flavor-scored by the AOCS-FNC had the following ratings: 4.0, 5.5, 6.8 and 8.0. These oils were examined by direct GC and combined GC/MS, and the profiles of volatiles obtained for three of them (the low, medium, and high scored oils) are shown in Figure 1. Chromatogram 1, which represents a high quality oil with a flavor score of 8, reveals few volatile components, in low concentrations. Pentane, the most prominent peak, is of moderate intensity. Two other peaks, hexanal and *trans*-2, *trans*-4-decadienal, which have been shown to be indicators of flavor quality in soybean oils (9), are of low intensity. This chromatogram is typical of that for a high quality, high flavor-scored soybean oil obtained by the direct GC method. Chromatogram 2, which was obtained from the soybean oil with a lower flavor score of 5.5, reflects a marked increase in the number and intensity of volatile components. In particular, the hexanal response has doubled, the *trans*-2,*trans*-4-decadienal has increased fourfold, and other components are present in significantly larger quantities. Chromatogram 3, which represents the poorest quality oil with a flavor score of 4, shows a dramatic increase in all volatile components, and certain flavor-related indicators as pentanal, hexanal, *trans*-2-heptenal, *trans*-2, *trans*-4-heptadienal, *trans*-2,*cis*-4-decadienal, and *trans*-2,*trans*-4-decadienal are conspicuously high. The comparison of chromatograms 2 and 3 is especially important—it reveals the sensitivity of the direct GC method of analysis over a flavor score range of only 1.5 units. It is not uncommon for flavor estimates by individual tasters to vary as much as three units for a given oil, whereas the direct GC method consistently detects subtle, yet reliable, differences in flavor quality over a much narrower range (9).

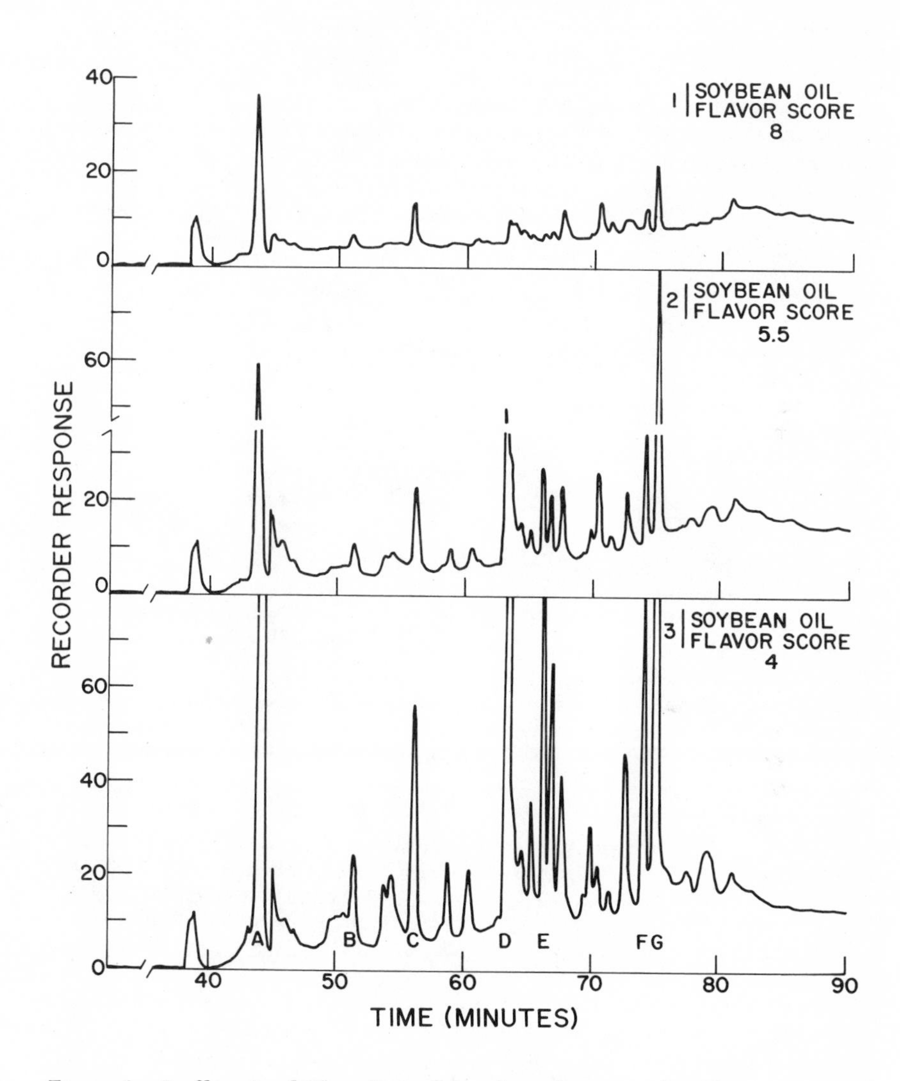

Figure 1. Profiles of volatiles obtained for three flavor-scored soybean oils. (A) pentane; (B) pentanal; (C) hexanal; (D) trans*-2-heptenal; (E)* trans*-2,*trans*-4-heptadienal; (F)* trans*-2,*cis*-4-decadienal; (G)* trans*-2,*trans*-4-decadienal*

The correlation coefficients between instrumental data and flavor scores for seven major volatile components detected in the flavor-scored oils are shown in Table I. All of the correlation coefficients are statistically significant at confidence levels of either 95% or 99%. These results agree with those obtained previously with soybean oils that had been progressively degraded by light exposure (9). The standard errors of estimate for the regressions of flavor score on the various peaks ranged from 0.11 to 0.57. The average standard errors of the mean for the 12 taste panels involved in this study ranged from 0.15 to

Table I
Regression Analysis of Soybean Oil
Flavor Scores with Log of Volatile Components

Volatiles	Correlation coefficient	F - Value	Standard error
Pentane	- 0.961	24.0*	0.57
Pentanal	- 0.977	41.2*	0.44
Hexanal	- 0.973	35.1*	0.48
t-2-Heptenal	- 0.999	743.7**	0.11
t,t-2,4-Heptadienal	- 0.991	108.7**	0.28
t,c-2,4-Decadienal	- 0.994	164.9**	0.23
t,t-2,4-Decadienal	- 0.991	105.1**	0.28
Total Volatiles	- 0.994	157.1**	0.23

*Significant at the 95% confidence level.
**Significant at the 99% confidence level.

0.62. Thus the direct GC procedure for estimating the flavor of soybean oil is at least as good, statistically, as a typical taste panel.

Table II compares taste panel flavor scores with those predicted by direct GC and combined GC/MS analysis. Use of the trans-2-heptenal peak area, the trans-2,trans-4-decadienal peak area, or the total volatiles area in the regression equation is equally satisfactory in predicting actual flavor scores.

Table II
Comparison of Oil Flavor Scores
Obtained by Taste Panels and Direct GC Analysis

Taste panel scores	Predicted scores based on instrumental data*		
	trans-2-heptenal peak	trans-2,trans-4-decadienal peak	Total volatiles
4.0	4.0	4.0	3.9
5.5	5.5	5.6	5.8
6.8	6.7	6.4	6.6
8.0	7.9	8.0	7.8

*Predicted from regression equation: a+b (log x).

Although this paper demonstrates the use of direct GC and combined GC/MS analysis in determining flavor quality for soybean oils, the method has extensive potential for application to other food products. It has proven effective for studying flavor changes associated with the stability of peanut butter (8,12). Neutral volatile components of mayonnaise have been detected, identified, and their flavor relationship examined (13). Volatile compounds present in various rice products, whole corn, and breakfast cereals were studied for their contribution to quality and flavor (14). Correlating volatile components of raw peanuts with flavor scores of the roasted product also appears to be quite attractive to plant geneticists for use in developing new peanut varieties (15).

Using the direct GC or combined GC/MS method of analysis with appropriate conditions should make it possible to obtain a profile of volatiles for most raw and processed food products, and to utilize such data for assessing flavor characteristics. Recent experiments with bacon samples have demonstrated the versatility of the method. Figure 2 shows the profile of volatiles obtained from commercial brands of bacon purchased from a local supermarket. In comparing the chromatograms for brands A, B, and C, no attempt is made to flavor-relate the products. Rather, they are presented merely to indicate the extent to which different peak components can be eluted and resolved by direct GC, and identified by combined GC/MS analysis. Brand C contains more volatile peaks of greater intensity than do brands A or B. Such compounds as acetic acid, 2-methyl-butanal, hexanal, and methoxyphenol are not pronounced in brands A and B, but are quite prominent in brand C. Similarly, an unidentified peak, eluting at approximately 32 min, is present in relatively high concentration in brand B and is less prominent in brands A and C. These observations indicate the scope and potential of direct GC and combined GC/MS analysis in detecting and identifying food volatiles.

When peaks responsible for flavor are established by taste panel procedures, their presence and intensity, or their absence in similar test products, can be determined with confidence. Preliminary experiments at The Center have indicated that nitrosamines can also be determined by combined GC/MS analysis. Further refinement of these initial experiments should make possible the simultaneous analysis of bacon or other meat samples for both flavor quality and nitrosamines. It is likely that efforts to increase the sensitivity and resolution of the direct GC and combined GC/MS methods of analysis will greatly enhance their utility in the broad area of flavor chemistry.

ABSTRACT

A simple, rapid and direct gas chromatographic technique elutes and resolves the volatile components from vegetable oils and relates them to flavor quality. The sample is placed in a

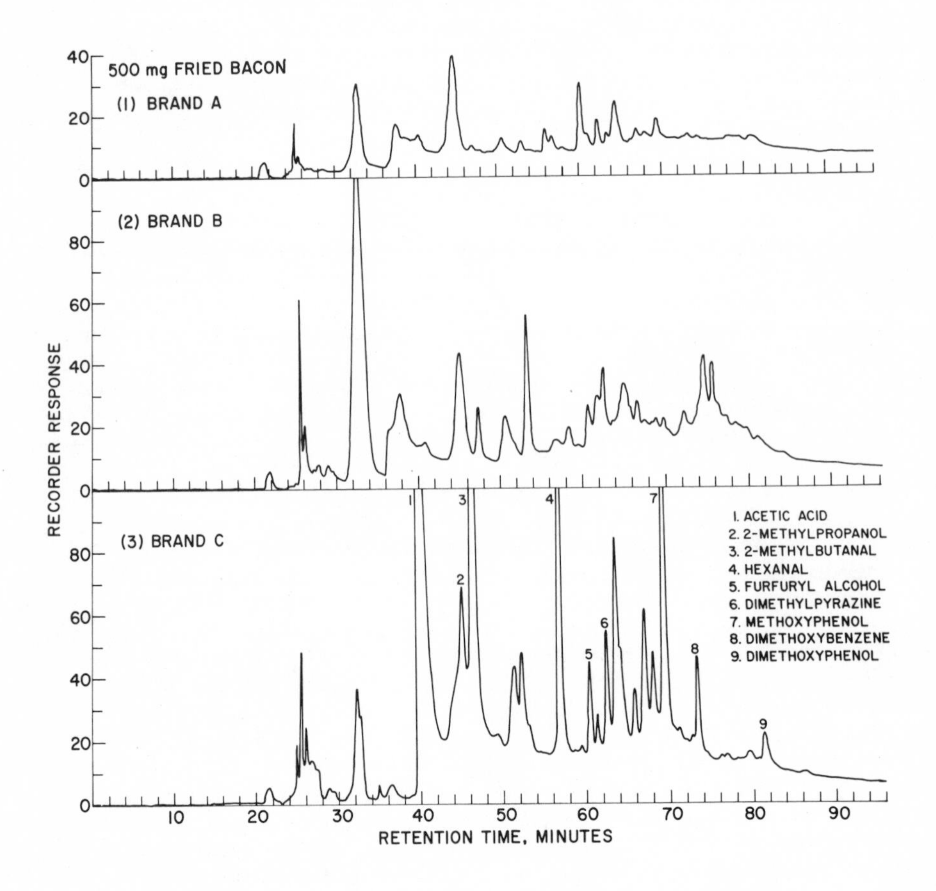

Figure 2. Profiles of volatiles obtained for three brands of bacon

gas chromatograph liner and secured in the heated injection port of the gas chromatograph, without prior enrichment of volatiles. Carrier gas, flowing through the heated oil, rapidly and efficiently elutes the volatiles, which are adsorbed on the relatively cool gas chromatographic column and subsequently resolved by temperature programming. The profile of volatiles obtained is an indication of the quality of oil flavor. Specific peaks of the chromatogram are identified by mass spectrometry. Correlation between taste panel flavor scores for the oils and the instrumental data obtained is significant at 99% and 95% confidence levels.

Literature Cited

1. Buttery, R. G., Teranishi, R., Anal. Chem. (1961) 33, 1439.
2. Teranishi, R., Buttery, R. G., Lundin, R. E., Anal. Chem. (1962) 34, 1033.
3. Smith, D. E., Coffman, J. R., Anal. Chem. (1960) 32, 1733.
4. Chang, S. S., J. Am. Oil Chem. Soc. (1961) 38, 669.
5. Lee, C. H., Swoboda, P. A. T., J. Sci. Food Agric. (1962) 13, 148.
6. Selke, E., Moser, H. A., Rohwedder, W. K., J. Am. Oil Chem. Soc. (1970) 47, 393.
7. Dupuy, H. P., Fore, S. P., Goldblatt, L. A., J. Am. Oil Chem. Soc. (1973) 50, 340.
8. Fore, S. P., Dupuy, H. P., Wadsworth, J. I., and L. A. Goldblatt, J. Am. Peanut Res. Educ. Assoc. (1973) 5, 59.
9. Dupuy, H. P., Rayner, E. T., Wadsworth, J. I., Legendre, M. G., J. Am. Oil Chem. Soc. (1977) 54, 445.
10. Williams, J. L., and Applewhite, T. H., J. Am. Oil Chem. Soc. (1977) 54, 461.
11. Jackson, H. W., and Giacherio, D. J., J. Am. Oil Chem. Soc. (1977) 54, 458.
12. Fore, S. P., Dupuy, H. P., and Wadsworth, J. I., Peanut Sci. (1976) 3 (2) 86.
13. Fore, S. P., Legendre, M. G., and Fisher, G. S., J. Am. Oil Chem. Soc. (1977), submitted for publication.
14. Legendre, M. G., Dupuy, H. P., Ory, R. L., and McIlrath, J. Agric. Food Chem. (1977), submitted for publication.
15. Brown, M. L., Wadsworth, J. I., Dupuy, H. P., and Mozingo, R. W., (1977) Peanut Sci. in press.

RECEIVED December 22, 1977

6

Chemical Changes Involved in the Oxidation of Lipids in Foods

D. A. LILLARD

Food Science Department, University of Georgia, Athens, GA 30602

The oxidation of lipids was recognized at least 150 years ago when Berzelius (1) described experiments illustrating the induction period, oxygen uptake, carbon dioxide formation and polymerization of linseed oil. Since that time there have been numerous research reports on lipid oxidation. The flavors that result from the oxidation of lipids have triggered a large number of these investigations. However, in addition to the formation of these flavorful secondary products, free radicals produced during lipid oxidation are capable of reacting with other constituents of food such as coloring substances, vitamins, enzymes, amino acids, and proteins. These reactions can have a deleterious effect on the nutritional quality of foods. The free radicals have also been implicated in reactions that lead to pathological changes in animal and human tissue (2).

This review will cover the chemical reactions involved in the oxidation of lipids in foods with emphasis on the formation of flavor compounds. It is by no means an inclusive review since many aspects of lipid oxidation have been covered by others (2, 3, 4, 5).

Mechanisms and Products of Lipid Oxidation

Unsaturated lipids are almost exclusively considered the initial substrate in lipid oxidation. The reaction is autocatalytic in that the oxidation products themselves catalyze the reaction and cause an increase in the reaction rate as oxidation proceeds. The universally accepted free radical reaction scheme for lipid oxidation can be written as follows:

Initiation

$$RH \longrightarrow R\cdot + H\cdot$$
$$RH + O_2 \longrightarrow ROO\cdot + H\cdot$$

0-8412-0418-7/78/47-075-068$05.00/0

Propagation

$R\cdot + O_2 \longrightarrow ROO\cdot$
$ROO\cdot + RH \longrightarrow ROOH + R\cdot$

Termination

$ROO\cdot + R\cdot \longrightarrow ROOR$
$R\cdot + R\cdot \longrightarrow R - R$
$ROO\cdot + ROO\cdot \longrightarrow ROOR + O_2$

where RH = unsaturated lipid
R· = lipid radical
ROO· = lipid peroxy radical

Once initiated, the oxidation reaction is also propagated by the breakdown of hydroperoxides to free radicals which in turn can accelerate the rate of lipid oxidation.

Initiation reaction. Although the free radical mechanism of lipid oxidation has been well established, the initial reaction or the formation of the first hydroperoxide, especially in hydroperoxide free lipids, is still open for debate. This is due to the fact that the reaction of lipids with O_2 to form free radicals or hydroperoxides is very unlikely. Hydroperoxide formation by this reaction would require a change in total electron spin since the hydroperoxides and lipids are in singlet states while oxygen is in the triplet state. The spin conservation of this improbable reaction would be satisfied and take place if singlet oxygen were the reactive species instead of ground state triplet oxygen (6). Using this hypothesis, Rawls et al. (6) proposed the following mechanism by which singlet oxygen could be formed by photo-chemical reactions in the presence of a sensitizer, and presented evidence that singlet oxygen reacts with lipids at a rate which is 1450 times faster than triplet oxygen:

$S + h\nu \longrightarrow {}^1S^* \longrightarrow {}^3S^*$

${}^3S^* + {}^3O_2 \longrightarrow {}^1O^*_2 + {}^1S$

${}^1O^*_2 + RH \longrightarrow ROOH$

$ROOH \longrightarrow$ free radicals

where

${}^1S$ = singlet state sensitizer
${}^1S^*$ = excited singlet state sensitizer
${}^3S^*$ = excited triplet state sensitizer
${}^3O_2$ = ground triplet state oxygen
${}^1O^*_2$ = excited singlet state oxygen

The sensitizers required to convert triplet oxygen to singlet oxygen are normally found in plant and animal tissues and consist of photosensitive compounds such as chlorophyll, pheophytin and myoglobin.

Recently, Aurand et al. (7) reported that singlet oxygen was the immediate source of the hydroperoxides that initiated milk lipid oxidation which was catalyzed by light, copper and xanthine oxidase. In light-induced oxidation, riboflavin served as the sensitizer by producing singlet oxygen directly from its photosensitized triplet state. In the copper and xanthine oxidase system, singlet oxygen was formed by dismutation of the superoxide anion which was formed in these systems. Oxidation was prevented in all three systems by the inclusion of a known singlet oxygen trapper (1,3-diphenylisobenzofuran) or a singlet oxygen quencher (1,4-diazabycyclo(2-2-2)octane). Lipid oxidation was also prevented in the copper and enzyme systems by addition of the superoxide dismutase enzyme. This enzyme catalyzes the superoxide dismutation to ground state oxygen thereby preventing the spontaneous dismutation of superoxide to singlet oxygen. This study provided evidence that singlet oxygen is the initial reactant with lipid when oxidized in systems other than those that are light-induced.

Although evidence is accumulating to indicate that singlet oxygen is a very likely initial reactant in several types of lipid oxidation, there is evidence that singlet oxygen does not participate in the oxidation of lipids that contain no sensitizers (8). Cort (8) found that the quenching agent, β-apo-8'-carotenal, had no effect on the oxidation of safflower oil, oleic acid and linoleic acid. In these experiments, no sensitizer was present and it was concluded that singlet oxygen was not involved in the air oxidation of these substrates. Recently, Terao and Matsushita (9) isolated the hydroperoxides from photosensitized oxidation of unsaturated fatty acid esters. They found that hydroperoxide groups of all isomers were attached to the carbon atoms which originally existed at both sides of a double bond and the double bond shifted to the adjacent positions. This distribution of hydroperoxide isomers was different from the distribution obtained from air oxidization of these fatty acid esters (10, 11). The photosensitized oxidation was inhibited by β-carotene (a singlet oxygen quencher) but was not inhibited by butyl hydroxytoluene (a free radical stopper). These results indicated that singlet oxygen oxidation differs from air oxidation.

It has yet to be determined if the initial reaction of all types of lipid oxidation involves singlet oxygen. A complete understanding of the singlet oxygen reaction with unsaturated lipids is desirable since most foods that are susceptible to oxidation contain components that are capable of inducing the formation of singlet oxygen. The use of effective nontoxic singlet oxygen quenchers in foods could prove to be very effective in increasing the stability of their unsaturated lipids (12).

Secondary reaction products. Lipid hydroperoxides are very unstable and break down to produce many types of secondary reaction products. This hydroperoxide decomposition proceeds by a free radical mechanism and can be illustrated by the following scheme (13):

$$\text{R-}\underset{\text{O-OH}}{\underset{|}{\text{CH}}}\text{-R} \longrightarrow \text{R-}\underset{\dot{\text{O}}}{\underset{|}{\text{CH}}}\text{-R} + \cdot\text{OH} \qquad \text{(A)}$$

$$\text{R-}\underset{\dot{\text{O}}}{\underset{|}{\text{CH}}}\text{-R} \longrightarrow \text{R-}\underset{\text{O}}{\underset{\|}{\text{CH}}} + \text{R}\cdot \qquad \text{(B)}$$

$$\text{R-}\underset{\dot{\text{O}}}{\underset{|}{\text{CH}}}\text{-R} + \text{R}^1\text{H} \longrightarrow \text{R-}\underset{\text{OH}}{\underset{|}{\text{C}}}\text{-R} + \text{R}^1\cdot \qquad \text{(C)}$$

$$\text{R-}\underset{\dot{\text{O}}}{\underset{|}{\text{CH}}}\text{-R} + \text{R}^1\cdot \longrightarrow \text{R-}\underset{\text{O}}{\underset{\|}{\text{C}}}\text{-R} + \text{R}^1\text{-H} \qquad \text{(D)}$$

$$\text{R-}\underset{\dot{\text{O}}}{\underset{|}{\text{CH}}}\text{-R} + \text{R}^1\text{O}\cdot \longrightarrow \text{R-}\underset{\text{O}}{\underset{\|}{\text{C}}}\text{-R} + \text{ROH} \qquad \text{(E)}$$

In reaction (A), the hydroperoxide is cleaved to alkoxy and hydroxy free radicals. Reactions (B-E) illustrate the reaction of the alkoxy free radical with other free radicals or molecules to form secondary products. Since hydroperoxides are flavorless (14), it is these secondary products that contribute to the oxidized flavor of food lipids. As is evident from the complex nature of this reaction and the complex composition of food lipids, numerous compounds are formed during the oxidation of lipids. The type of oxidation products include carbonyl compounds, alcohols, semi-aldehydes, acids, hydrocarbons, lactones and esters (13, 15).

Many of the compounds identified as oxidation products are produced by the oxidation of the primary scission products (15, 16, 17, 18).

Lillard and Day (15) made one of the first systematic studies of the oxidative breakdown of initially formed carbonyl products. They found that the rate of autoxidation depended upon the class of carbonyl being oxidized. When oxidized at 45°C in an oxygen atmosphere, n-nonanal had an induction period of 12 hours and the only oxidation product was n-nonanoic acid. No induction period was observed for non-2-enal and hepta-2,4-dienal. The degradation products for non-2-enal were ethanal, n-heptanal, n-octanal, propanal, α-ketooctanal, glyoxal, α-ketononanal and α-ketoheptanal. The carbonyl compounds identified from oxidized hepta-2,4-dienal were propanal, ethanal, n-butanal, cis-but-1-en-1,4-dial, α -ketopentanal, glyoxal, α -ketoheptanal and α-keto-hexanal. Oct-1-en-3-one did not oxidize when held at 45°C for

52 hours. Recently, Michalski and Hammond (17) confirmed the work of Lillard and Day by using radio-tracer techniques to follow the oxidation of similar classes of carbonyl compounds in oxidizing soybean oil. Also, other studies have shown that n-alkanals can be oxidized under certain conditions to produce aldehydes, alcohols, esters, hydrocarbons and lactones (16, 18). These investigations have illustrated that continued oxidation of the initial secondary products can account for many of the compounds found in oxidized lipids whose origin could not be attributed to the cleavage of the hydroperoxides believed to be present in the lipids under study.

Factors Affecting Lipid Oxidation

Lipid composition and structure. In model systems consisting of pure fatty acid esters, resistance to oxidation can be related to the ease with which the initiation reaction can occur. The hydrogen lability of the methylene carbons on which the free radicals are formed can be grouped according to the number and type of unsaturated bonds in the fatty acid molecule (19). However, lipids in food are not simple, pure components and are in a close contact with other oxidizable compounds, enzymes, and various types of prooxidants and antioxidants. For these reasons, the oxidative stability of foods is not always related to the degree and type of unsaturated lipids they contain (20, 21).

In order to determine if the oxidative stability of an oil is related to the fatty acid composition, Graboski (21) prepared oil blends composed of sunflower seed oil-olive oil, sunflower seed oil-linseed oil and olive oil-linseed oil to obtain samples of various degrees of unsaturation. These oil blends were checked for oxidative stability using the active oxygen method (AOM). There was no correlation between the individual fatty acid content of these oil blends and oxidative stability.

Using oxidation rate factors of individual fatty acids (22) as correction factors, the ratio of total unsaturated fatty acids to the saturated fatty acids in each oil blend was calculated according to the following equation:

$$R = \frac{11^{*}(C18:1) + 114^{*}(C18:2) + 179^{*}(C18:3)}{1(18)}$$

* = oxidation rate factors proposed by Stirton *et al.* (22)

( ) = concentration of fatty acids

The logarithm of this ratio gave a correlation of -.83 with the oxidative stability of the oil. This suggested that the oxidation factors could be used in conjunction with fatty acid composition of the oil to predict the oxidative stability of the oil. However, when these factors were applied to the fatty acid composition of five sunflower seed oils, no correlation was obtained with the oxidative stability of the oils (21). Perhaps the effect of prooxidants and antioxidants in the oils had more influence on the rate of oxidation than did differences in fatty acid composition of the oils.

Raghuveer and Hammond (23) found that the position of the unsaturated fatty acid in the triglyceride could influence the oxidation rate of the oil. An oil would be more stable to oxidation if more of the unsaturated fatty acids were located in the 2-position of triglycerides than if they were located in the 1 and 3 positions. Recent work at Georgia has shown that the oxidative stability of peanut oil is not related to the amount of linoleic in the 2-position of the triglycerides (24).

Oxidation catalysts. Heavy metals, primarily those having two valency states with suitable oxidation-reduction potentials, increase the rate of lipid oxidation. They may be effective as secondary catalysts of oxidation where they act as electron donors to hydroperoxides to produce RO· free radicals (25). The M^{+3} can be converted back to M^{+2} by reacting with hydroperoxides to form ROO· radicals.

$$ROOH + M^{+2} \longrightarrow RO\cdot + M^{+3} + OH^-$$

$$ROOH + M^{+3} \longrightarrow ROO\cdot + M^{+2} + H^+$$

These two reactions can establish additional chain sequences in the oxidizing lipid. There is evidence, however, that metals may be involved as primary catalysts and can initiate lipid oxidation (26) as discussed previously. Aurand et al. (13) illustrated that Cu^{+2} catalyzed milk took place by the formation of singlet oxygen. The source of singlet oxygen was the dismutation of superoxide anion which was formed by the action of Cu^{+2} on oxygen.

The impact of heme pigments on lipid oxidation has been reviewed recently (4, 27, 28). It is a known fact that heme proteins can catalyze lipid oxidation but the question yet to be answered satisfactorily is the exact state of the heme iron when catalysis of lipid oxidation occurs (27). Several investigators (29, 30) believe that the iron of heme groups must be in the oxidized form (Fe^{+3}) in order to function as a catalyst of lipid oxidation while other workers (31, 32) indicate that both Fe^{+2} and Fe^{+3} forms are equally active as oxidation catalysts. Oxidation of myoglobin to metmyoglobin has been shown to accompany lipid oxidation in model systems (33). However, in this case, was it necessary for the myoglobin to be in the oxidized form in order to be an oxidation catalyst or was metmyoglobin formed as a scission product of a myoglobin-lipid complex.

Antioxidants. Chemicals when added to foods to prevent or delay the onset of lipid oxidation are called antioxidants. Phenolic type compounds such as butylated hydroxy anisole (BHA), butylated hydroxy toluene (BHT) and tocopherol can act as free radical stoppers by donating hydrogen to the free radicals. Chelating agents such as ethylenediaminetetracetic acid and citric acid can be classified as free-radical-production preventers by tying up metal catalysts (4). Certain compounds such as ascorbic acid have the ability to provide a synergistic effect when used with phenolic type antioxidants. This synergistic action of ascorbic acid is attributed to its ability to regenerate antioxidants by supplying hydrogen to the phenoxy radical (34) or to its ability to function as an oxygen scavenger in some systems (8). Cort (8) reported that ascorbyl palmitate, an ester of ascorbic acid, has better solvent properties than ascorbic acid and is more effective at 0.01% concentration than BHT or BHA at 0.02% concentrations in delaying the oxidation of peanut, safflower, sunflower and corn oils.

Tocopherols are classified among the natural antioxidants. Although they have been shown to be effective in animal fats, the activity of tocopherols added to vegetable oils has always been low. The reason usually given for this has been that tocopherols naturally present in vegetable oils mask additional activity. However, Cort (8), using molecular distillation, removed tocopherols from soybean and safflower oil and found that the addition of tocopherols did not effectively improve the stability of the oils. After demonstrating that tocopherols were more active in oleic acid than in linoleic acid, and that tocopherol activity in animal fats was similar to the activity in oleic acid, Cort concluded that the variations in tocopherols activity in different substrates may be due to the type of unsaturated fatty acids in these substrates.

Carlsson *et al.* (12) demonstrated that free radical scavenger type antioxidants had no influence on the formation of hydroperoxide in oils oxidized by near UV and visible light. Nickel (II) chelates, which are singlet oxygen quenchers, retarded lipid deterioration under their experimental conditions.

Antioxidants often exhibit variable degrees of efficiency in protecting food systems. They usually give excellent protection in unsaturated fats and oils but only moderate protection in most other foods.

Lipid Oxidation in Various Flesh Foods

As indicated earlier, most studies on mechanisms of lipid oxidation have been done on model systems consisting of unsaturated fatty acids or their esters. Although it may be assumed that the same type of reactions occur in food systems, due to the complex nature of foods many types of additional interactions and reactions may occur. In this section, several types of flesh foods and their oxidative deteriorations will be discussed.

Although animal lipids are considered to be fairly saturated, sufficient amounts of unsaturated lipids are found in the phospholipid fraction of the intramuscular lipids to produce oxidized flavors in meat products (35). Furthermore, these phospholipids are in close contact with various oxidation catalysts that exist in muscle tissue. Recently, emphasis has been placed on increasing the amount of polyunsaturated fatty acids in ruminant fats. This can be accomplished by encapsulating polyunsaturated oil droplets with a layer of protein (casein) which is treated with formaldehyde to prevent hydrogenation of the unsaturated fatty acids by the microflora in the rumen (36). This procedure increases the amount of unsaturated fats in meats. This increase may be desirable for nutritional reasons but may lead to increased problems of oxidative deterioration upon storage. Scientists at the Eastern Regional Research Center (37, 38) studied the effect of protected safflower oil on the linoleic acid content and oxidative stability of rendered fat from cattle of different ages that were fed the oil. They reported that the depot fat of all animals on protected diets contained higher amounts of linoleic acid than depot fats from control animals. The magnitude of the increase was influenced by the age of the animal and the amount of protected safflower oil in the feed. In the growing and mature steer groups, the increased linoleic acid levels decreased the stability of the rendered fats. There did not appear to be a relationship between deposition of tocopherol in depot fat and protected safflower diets when all animals received a 20 mg d-α-tocopheryl acetate per day. In experiments using young calves, an increased linoleic acid content in depot fat was also obtained in animals fed protected safflower oil. However, high increases in linoleic acid did not result in decreased stability of the fat due to high tocopherol levels in the fat (38). Bremner *et al.* (39) reported on the oxidative changes during frozen storage of meat (mutton, lamb, beef) with high linoleic acid content. Peroxides occurred at a faster rate in adipose tissue from high linoleic meat than conventional meat when stored at -10°C. Taste panel evaluations indicated that high linoleic meat stored at -10°C developed rancid odors and flavors 2-3 times more rapidly than conventional meat. At -20°C, the rate of peroxide formation in high linoleic meat was greatly reduced but the rate was still equal to that of conventional meat stored at -10°C. If production of high linoleic meat becomes commercially important, conventional methods of packaging and storing frozen meats may not be adequate for this type of product and alternate methods should be sought.

Perhaps the most serious flavor defect attributed to the oxidation of lipid in meat is the warmed-over-flavor (WOF). This descriptive term was used by Timms and Watts (40) to describe the rapid development of oxidative rancidity in cooked meat during short-term refrigerated storage. Originally, most research indicated that the WOF was due to metmyoglobin-catalyzed lipid oxidation and that the cooking process denatured myoglobin and this increased the catalytic activity of the heme proteins (30, 32, 41). Recently, it has been shown that non-heme iron may be more important than heme iron as a prooxidant in cooked meat (42, 43). Wilson et al. (44) demonstrated that turkey meat was most susceptible to WOF development, followed by chicken, pork, beef and mutton. Analysis of these data revealed that unsaturated fatty acids of the phospholipids may be the substrate being oxidized in all the meats except pork. In pork, the total lipid levels seem to be the major contributor to WOF.

In cured meats, the lower level of lipid oxidation observed has been attributed to the ability of the curing agent, sodium nitrite, to form a complex with the heme proteins thereby maintaining the heme iron in the ferrous form. The reduced heme is postulated to be a slower oxidation catalyst than the oxidized heme. Upon storage, the cured meat pigment is converted to the ferric form which results in increased oxidation of the lipids (30). Greene and Price (27) reported that increases in lipid oxidation, as measured by TBA tests, had no effect on cured meat flavor as scored by a taste panel. Possibly some of the products of lipid oxidation contribute to the cured meat flavor. This may be especially true in cured and aged pork products such as country-style hams. Lillard and Ayres (45) reported that many of the carbonyl compounds in country-style hams are the same as the compounds identified in oxidized pork lipids.

Although oxidation of fresh fish tissue has not been considered a major cause of flavor deterioration because microbial spoilage sets in first (46), some fish do contain sufficient amounts of unsaturated lipids to undergo rapid oxidation when stored either fresh or frozen (47, 48, 49, 50). Recently, Ke and co-workers found that lipids associated with the skin of mackerel were oxidized at a rate eight times faster than dark and light tissue lipids and concluded that this was due to one or more fat soluble prooxidants located in the skin (48). Other oxidative catalysts found in fish appear to be similar to those found in other flesh foods; their concentrations vary within muscle tissues of the same fish as well as within species of fish, location, and time of year harvested (4). Deng and co-workers (50) have retarded the onset of oxidation in frozen mullet by use of ascorbic acid and mono-tertiary butylhydroquinone in combination with vacuum packaging. Ascorbic acid was found to act as an antioxidant or prooxidant depending on the concentration used (51). In dark flesh of mullet, ascobic acid acted as an antioxidant at concentrations above 500 ppm and a prooxidant at concentrations below 500 ppm. The antioxidant to prooxidant shift was observed at 50 ppm in the white flesh.

The use of machines to remove flesh from bones of poultry and fish has created a source of proteins which can be used in emulsified and processed food products. However, this protein has limited use for food ingredients because of flavor instability during storage. Lipid oxidation has been considered the major cause of this flavor deterioration (52, 53, 54, 55, 56, 57, 58). Moerck and Ball (55) reported that highly unsaturated fatty acids in the phospholipids were the source of the oxidative attack in mechanically deboned poultry (MDP). Lee et al. (56) concluded that the hemoproteins were the predominant catalysts of lipid oxidation in MDP and that the relative concentration ratio of polyunsaturated fatty acids to hemoprotein was in the range where heme-catalyzed oxidation would occur close to the maximum rate.

Lee and Toledo (58) investigated lipid oxidation in mullet flesh during mechanical deboning and storage. They illustrated that use of non-stainless steel equipment decreased the oxidative stability of the fish flesh probably due to an increase in the iron content of the flesh. Washing the deboned flesh before storage increased oxidative stability when stored at -16°C but had no effect when stored at 3°C. They also indicated that red muscle was the most susceptible to oxidation possibly due to the effect of heme pigments or a greater concentration of unsaturated lipids in this tissue. Silberstein and Lillard (59) recently characterized the heme pigments in phosphate buffer extracts of hand and mechanically deboned mullet. Mechanical deboning increased the concentration of hemoglobin in the deboned mullet but had very little influence on the myoglobin content. Furthermore, some hand deboned samples contained as much heme protein as the mechanically deboned samples. This variability in the amount of heme protein in fish was also reported by Brown (60) in his study on heme proteins in tuna fish. Oxidation studies using oleic acid as a substrate, revealed that the catalytic activity of the extracts was dependent on their total heme protein content (59). Using purified hemoglobin and myoglobin as catalysts, Silberstein and Lillard (59) found that myoglobin had a higher catalytic effect than hemoglobin and that samples containing the same amount of heme protein but different ratios of myoglobin to hemoglobin had increased rates of oxygen uptake as the ratio of myoglobin to hemoglobin increased. This indicates that the myoglobin to hemoglobin ratio as well as the total concentration of heme protein should be considered when determining the role of heme pigments as prooxidants in food systems.

Fischer and Deng (61) reported heme iron as the major catalyst in homogenates of the dark flesh of mullet. They did not determine the amount of hemoglobin and myoglobin in the extracts, but they did indicate a high percentage of non-heme iron in the dark muscle of mullet. They suggested that the non-heme iron should be characterized and its role in oxidative stability be determined.

Conclusions

A great deal of research has been done in the area of lipid oxidation of foods and a great deal is known about this complex series of reactions. However, additional information is needed in order to eliminate this problem in foods. For instance, is the singlet oxygen theory the answer to the initiation reaction in all oxidation systems? I doubt that it is. A better understanding of its implication in lipid oxidation is needed and may result in better control of some types of lipid oxidation. Also, if the trend to produce more unsaturated meats and milk continues for nutritional reasons, conventional processing and storage methods of these products will have to be reevaluated in terms of their effect on the oxidative stability of these new products. The oxidative stability of foods is still a problem and continued research in this area is still needed.

Literature Cited

1. Berzelius, J. J., "Larbok i kemien. IV.", Henry A. Nordstrom, Stockholm (1827).
2. Barber, A. A., Bernhein, F., Adv. Gerontol Res. (1967)2, 355.
3. Schultz, H. W., Day, E. A., Sinnhuber, R. O., "Lipids and their oxidation", 423, Avi Pub. Co., Westport, Conn. (1962).
4. Labuza, T. P., Critical Rv. Food Tech. (1971)2, 355.
5. Loury, M., Lipids (1972)7, 671.
6. Rawls, H. R., van Santen, P. J., J. Amer. Oil Chem. Soc. (1970)47, 121.
7. Aurand, L. W., Boone, N. H., Giddings, G. G., J. Dairy Sci. (1977)60, 363.
8. Cort, W. M., J. Amer. Oil Chem. Soc. (1974)51, 321.
9. Terao, J., Matsushita, S., J. Amer. Oil Chem. Soc. (1977) 54, 234.
10. Lundberg, W. O., Jarvi, P., in "Progress in the Chemistry of Fats and Other Lipids," p. 377, ed. R. T. Holman, Pergamon Press, New York (1971) 9.
11. Frankel, F. N., in "Lipids and Their Oxidation," p. 51, eds. H. W. Schultz, E. A. Day, R. O. Sinnhuber, Avi Pub. Co., Westport, Conn. (1962).
12. Carlsson, D. J., Suprunchuk, T., Wiles, D. M., J. Amer. Oil Chem. Soc. (1976)53, 656.
13. Keeney, M., in "Lipids and Their Oxidation," p. 79, eds. H. W. Schultz, E. A. Day, R. O. Sinnhuber, Avi Pub. Co., Westport, Conn. (1962).
14. Lea, C. H., Sawobada, Chem. and Ind. (1958)46, 1289.
15. Lillard, D. A., Day, E. A., J. Amer. Oil Chem. Soc. (1964) 41, 549.
16. Loury, M., Forney, M., Rev. Franc. Corps Gras (1968)15, 663.
17. Michalski, S. T., Hammond, E. G., J. Amer. Oil Che. Soc. (1972)49, 563.
18. Palamand, S. R., Dieckmann, R. H., J. Agr. Food Chem. (1974) 22, 503.

19. Farmer, E. H., Koch, H. P., Sutton, D. A., J. Chem. Soc. (1943) 541.
20. Sattar, A., Deman, J. M., J. Amer. Oil Chem. Soc. (1976) 53, 473.
21. Graboski, T. A., "The Isolation and Identification of the Fatty Acids of Phospholipids of Sunflower Seed (Helianthus annuus) Oil", Ph.D. Thesis, University of Georgia, Athens, Georgia (1974).
22. Stirton, A. J., Turner, J., Reimenschneider, R. W., Oil and Soap (1945) 22, 81.
23. Raghuveer, K. G., Hammond, E. G., J. Amer. Oil Chem. Soc. (1967) 44, 239.
24. Hokes, J., "Factors Affecting the Oxidative Stability of Oils from Various Peanut Cultivars", M.S. Thesis, University of Georgia, Athens, Georgia (1977).
25. Ingold, K. U. in "Lipids and their Oxidation", p. 93, eds. H. W. Schultz, E. A. Day, R. O. Sinnhuber, Avi Pub. Co., Westport, Conn. (1962).
26. Heaton, M. W., Uri, N., J. Lipid Res. (1961) 2, 152.
27. Greene, B. E., Price, L. G., J. Agr. Food Chem. (1975) 23, 164.
28. Greene, B. E., J. Amer. Oil Chem. Soc. (1971) 48, 637.
29. Greene, B. E., J. Food Sci., (1969) 34, 110.
30. Younathan, M. T., Watts, B. M., Food Res. (1960) 25, 538.
31. Brown, W. D., Harris, L. S., Olcott, H. S., Arch. Biochem. Biophys. (1963) 101, 14.
32. Hirano, Y., Olcott, H. S., J. Amer. Oil Chem. Soc. (1971) 48, 523.
33. Koizumi, C., Nonaka, J., Brown, W. D., J. Food Sci. (1973) 38, 813.
34. Bauerfeind, J. C., Pinkert, D. M., in "Advances in Food Research," eds. C. O. Chichester, E. M. Mrak, G. F. Stewart, Academic Press, New York (1970) p. 219.
35. Hornstein, R. T., Crowe, P. F., Heimberg, M. J., J. Food Sci. (1961) 26, 581.
36. Scott, T. W., Cook, L. J., Mills, S. C., J. Amer. Oil Chem. Soc. (1971) 48, 358.
37. Ellis, R., Kimoto, W. I., Bitman, J., Edmondson, L. F., J. Amer. Oil Chem. Soc. (1974) 51, 4.
38. Kimoto, W. I., Ellis, R., Wasserman, A. E., Oltjen, R., J. Amer. Oil Chem. Soc. (1974) 51, 401.
39. Bremner, H. A., Ford, A. L., Macfarlane, J. J., Ratcliff, N. T., J. Food Sci. (1976) 41, 757.
40. Timms, M. J., Watts, B. M., Food Technol. (1958) 12, 240.
41. Kendrick, J., Watts, B. M., Lipids (1969) 4, 454.
42. Sato, K., Hegarty, G. R., J. Food Sci. (1971) 36, 1098.
43. Love, J. D., Pearson, A. M., J. Agr. Food Chem. (1974) 22, 1032.
44. Wilson, B. R., Pearson, A. M., Shorland, F. B., J. Agr. Food Chem. (1976) 24, 7.

45. Lillard, D. A., Ayres, J. C., Food Technol. (1969) 23, 117.
46. Olcott, H. S., in "Lipids and their oxidation," p. 173, eds. H. W. Schultz, E. A. Day, R. O. Sinnhuber, Avi Pub. Co., Westport, Conn. (1962).
47. Mendenhall, V. T., J. Food Sci. (1972) 37, 547.
48. Ke, P. J., Ackman, R. G., J. Amer. Oil Chem. Soc. (1976) 53, 636.
49. Ke, P. J., Nash, D. M., Ackman, R. G., Inst. Can. Sci. Technol. J. (1976) 9, 136.
50. Deng, J. C., Matthews, R. F., Watson, C. M., J. Food Sci., (1977) 42, 344.
51. Deng, J. C., Inst. Food Technol. 37th Annual Meeting (1977) Philadelphia, Pa., June 5-8.
52. Maxon, S. T., Marion, W. W., Poultry Sci. (1970) 49, 1412.
53. Dimick, P. S., MacNeil, J. H., Grunden, L. P., J. Food Sci., (1972) 37, 974.
54. Froning, G. W., Arnold, R. G., Mandigo, R. W., Neth, C. E., Hartung, T. E., J. Food Sci. (1971) 36, 974.
55. Moerck, K. E., Ball, H. R. Jr., J. Food Sci. (1974) 39, 876.
56. Lee, Y. B., Hargus, G. L., Kirkpatrick, J. A., Berner, D. L., Forsythe, R. H., J. Food Sci. (1975) 40, 964.
57. Babbitt, J. K., Law, D. K., Crawford, D. L., J. Food Sci. (1976) 41, 35.
58. Lee, C. M., Toledo, R. T., J. Food Sci. (1977) 42, IN PRESS.
59. Silberstein, D. A., Lillard, D. A., Inst. Food Technol. 37th Annual Meeting (1977) Philadelphia, Pa., June 5-8.
60. Brown, W. D., J. Food Sci. (1962) 27, 26.
61. Fischer, J., Deng, J. C., J. Food Sci. (1977) 42, 610.

RECEIVED December 22, 1977

7

Flavor Problems in the Usage of Soybean Oil and Meal

H. J. DUTTON

Northern Regional Research Center, Federal Research, Science and Education Administration, U.S. Department of Agriculture, Peoria, IL 61604

Oil and meal flavors in soybeans appear to present a bifurcated subject. But as we shall see, the problems of the meal are in part problems of the oil. Since residual lipids of soybean flakes constitute the precursors for odors and flavors, knowledge of the deteriorative reactions of fats is basic to the understanding of flavor development in soybean meal products. In this review the emphasis has been placed on lipid-derived flavors and their precursors, degradative reactions, separations, analyses, and psychometric evaluations.

To set the discussion of soybean oil flavor in proper perspective, one might first turn to the recent statistics on the disposition of soybean oil. Margarine and shortening comprise large outlets for soybean oil. Cooking and salad oils have been taking an increasing proportion of the "pie" probably because of the relationship of polyunsaturation to blood cholesterol lowering. With this magnitude of consumption of soybean oil the question must inevitably raise in the reader's mind, "What is the importance of the flavor problem?" (Figure 1)

In the early 1940's soybean oil was considered neither a good industrial paint oil, it was slow to dry, nor a good edible oil. In those days soybean oil flavor was considered the "Number One Problem of the Soybean Industry." Only under the exigencies of World War II was it added to margarines--and then to the absolute limit of 30%! The history of soybean oil is a story of progress from a minor edible oil of dubious value in the 1940's to a major edible oil proudly labeled on premium products of the 1970's. It is a story of cooperation between government research on the one hand and industrial implementations of research findings on the other.

Trivial as it may seem at this time, the first significant milestone in research was the development of a more objective method of assessing flavor and odor (1). With this procedure, numerical values for flavor intensity from a taste panel in one plant could be reproduced quite easily by a panel in another company or research institution. Equally important perhaps,

0-8412-0418-7/78/47-075-081$05.00/0

Table I
Milestones in Improving Flavor Stability of Soybean Oil

Date	NRRC Research	Industry Response
1945	Standardized taste test	Worldwide acceptance
1945	Trace metals	Brass valves, sheet steel out
1948	Metal deactivators	"Nary a lb without citric acid"
1948	Flavor is oxidation	Inert gas blanketing
1951	Precursor--linolenic	
	--breed it out	Homozygous (it can't be done)
	--extract it out	Practiced but now obsolete
	--hydrogenate it out	"Specially processed soybean oil"
1966	Recognition of room odor problem	
1966	Copper catalysts	
1974		Commercial production of cooking oils by copper catalysts

research finally had a reliable way of comparing samples and assessing more reliably the benefit of a given processing treatment rather than relying on the judgment of a single "expert."

With this new tool for evaluation, trace metals were identified as having special significance in the flavor stability of soybean oil compared to other edible fats and oils; whereas, cottonseed oil can tolerate copper and iron in the parts per million (ppm) range, soybean oil is ruined by as little as 0.3 ppm of iron and 0.01 ppm of copper (2). What followed the discovery of the deleterious effect of trace metals, especially in soybean oil, was removal of brass valves in oil refineries and conversion from cold rolled steel deodorizers to stainless-steel and even to nickel.

Strange as it may seem in retrospect, scientists had to establish that "soybean flavor reversion," as it was incorrectly called, was an oxidative process. When we sharpened up our analytical tools, the relation of peroxidation to off-flavor became unmistakable. The response of industry to the conclusion that "reversion is oxidation" was to blanket oils with inert gas at all critical high-temperature steps, including final packaging.

The next milestone has the aspects of a cloak-and-dagger story. At the close of World War II, Mr. Warren H. Goss, a chemical engineer at Northern Regional Research Center (NRRC), was commissioned a major in the Army with special assignment to follow Patton's advancing tanks through Germany and to investigate the German oilseed industry. As the troops advanced, he kept hearing about a recipe to cure soybean "reversion," but not until he reached Hamburg did he learn exact details. It was a strange formula involving many washings...such as contacting oil with water glass; but weird or not, when tested at NRRC it worked. It worked, as we were to learn, not because of the unusual washing treatments, but because citric acid was added to the deodorizer (3). Citric acid, we were to learn at NRRC, functioned by binding or complexing the deleterious traces of prooxidant metals. Based upon this discovery came the surge of metal deactivators--i.e, sorbitol, phosphoric acid, lecithin, polycarboxy acids, and starch phosphates. The immediate response of industry was to adopt metal deactivation, and I suspect that today there is not a pound of soybean oil product not protected by citric acid or some other metal deactivator.

These palliative steps, important as they were, still begged the question as to what causes off-flavor to develop--i.e., what are their unstable precursors? Unsaponifiables, i.e., sterols, were suspect. Circumstantial evidence pointed to the 7% content of linolenic acid, which draws its name from linseed oil where this trienoic fatty acid amounts to ca. 50%.

In what is now a classic experiment, 9% linolenic acid was interesterified into the glyceride structure of a nonreverting nonlinolenic acid oil; namely, cottonseed oil. The taste panel identified cottonseed oil interesterified with linolenic acid as soybean oil (4)!

Armed with this new basic information, what could be done? Three alternatives suggest themselves with regard to linolenic acid removal: (1) Breed it out; (2) extract it out; or (3) react it out.

Of the three alternatives listed, reacting out linolenic acid was chosen as the most practical research approach--and thereupon began a long search for selective hydrogenation catalysts--those that would react with linolenic acid but not attack the desired, essential polyunsaturated fatty acid--linoleic acid.

Fortunately, at this time our basic researches of catalyst selectivity bore fruit. NRRC's scientists found that among many metals active as hydrogenation catalysts, copper behaved with almost enzymatic specificity, hydrogenating linolenic acid some 15 to 20 times more rapidly than linoleic acid (5). It meant that not 3 to 4% linolenic acid in salad oils characteristic of nickel hydrogenation of soybean oil but "zero" percent linolenic oils could be produced, with little attack on the essential linoleic acid and with concomitantly low winterization losses. Room odor studies, conducted by our taste-odor panel, could scarcely detect the fishy odors characteristic of unhydrogenated soybean oil or of soybean oil partially hydrogenated by conventional nickel catalysts (6).

Now, it can be reported that in the United States a new plant has been built and has come on stream using copper catalysts. A large producer in Europe was observed to be test marketing three brands of copper-hydrogenated soybean oil in France, and a major French oil processor has a plant operating with copper catalysts.

The special interest in copper-hydrogenated soybean oil in France stems in part from the quadrupling in price of peanut oil coming from Africa. (Aflatoxin in the meal lowers its feed value and therefore raises oil prices.) A French law permits soybean oil with less than 2% linolenic acid to be sold as a salad or cooking oil. The copper-hydrogenation soybean oil may rank as one of the important developments in edible oil production in recent years.

Although samples of copper-hydrogenated soybean oil produced abroad have appeared to our taste panel to be of the same quality as that which is successful in the United States, the French housewife, not the taste expert, determines the consumer acceptance. Because the copper-hydrogenated soybean oil when stored or heated does not have the odor or flavor of the familiar peanut oil, there is some question as to whether the copper-hydrogenated soybean oil can replace it. Likewise, in the Mediterranean region where olive oil is the traditionally used deep-fat frying fat, the future of edible soybean oil will be determined by its acceptability in odor and flavor upon cooking. If soybean oil is to find new markets in France and in the Mediterranean region, it must meet this remaining problem of room odor.

The odor principles that cause the residual odor from deep-fat frying fats then assume practical importance. One method of studying them is the gas chromatograph-mass spectrometer system combined with "nose in the computer loop" as shown in Figure 2. In what has been dubbed a "micro frying pan" either the oil or individual purified constituents of oil are heated to deep-fat frying temperatures, while air is passed over the surface and then through a needle downward into a gas chromatographic column. The volatile odors are frozen out in the first few inches of this column at dry ice temperatures. After 5 minutes of collection of the odors, the gas chromatograph is returned to its usual source of gas pressure, and the chromatograph is programmed. The effluent from the column is split in three ways. The flame ionization detector, of course, provides us with the information on how many compounds and how much of each. The second split goes to the mass spectrometer-computer system which answers the question as to what the component is. And finally, a third part is sent for sniffing by the human nose and for its owner's recording the intensity of the odor sensation by means by a voltage dividing potentiometer and recorder. Thus, he simultaneously prepares and draws his intensiogram at the same time the gas chromatograph is recording the chromatogram. Whereas a flame ionization detector tells how many compounds and how much, the mass spectrometer tells what they are, the human nose tells how significant is the odor.

An example of the application of this technique to the room odor and the volatiles of heated soybean oil is given in Figure 3. The lower part is the usual gas chromatogram and the upper section is that of an intensiogram as drawn by the human observer (7). As

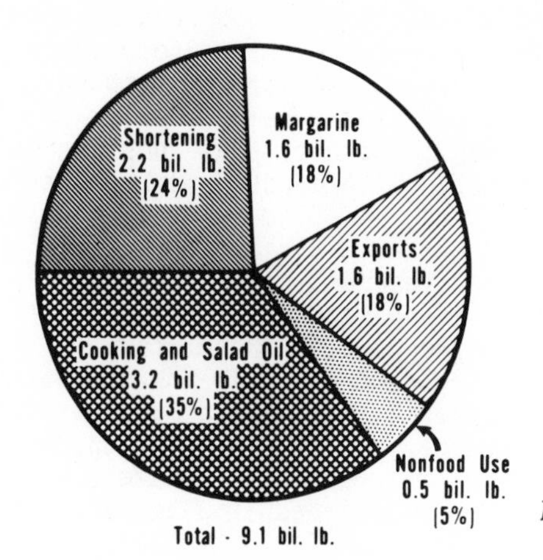

Figure 1. Disposition of U.S. soybean oil—1976

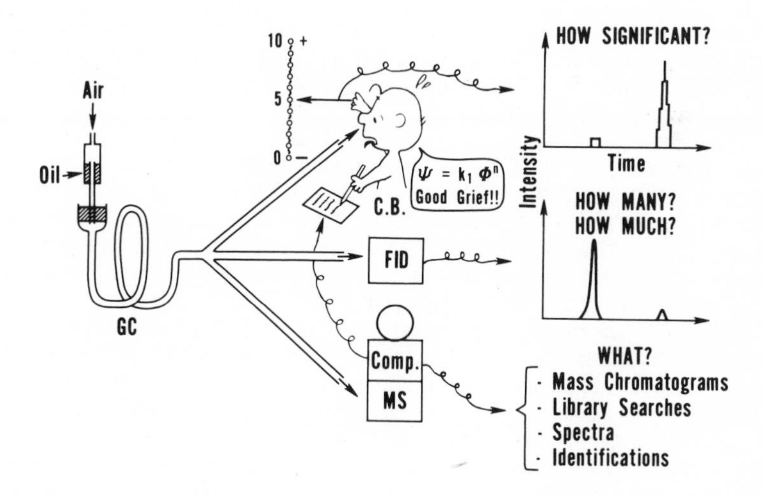

Figure 2. A nose in the GC-computer loop

would be expected, there are large peaks on the gas chromatogram that have small odor response and, contrariwise, there are minimal peaks in the gas chromatogram that give a large olfactory response. At the present time we are not only able to identify most of the peaks of this chromatogram chemically, but we have also made similar measurements upon the purified components that comprise soybean oil and are thus able to associate individual peaks of this curve as having their precursor of oleic, linoleic, linolenic, or saturated acids.

Unfortunately, it is widely understood but rarely acknowledged that the total volatiles collected at the exit port of a gas chromatograph should not differ from the odor of the mixture injected. Specifically, the whole should be equal to the sum of its parts. Because of the heat lability of odor principles with which we deal and the necessary temperatures for their volatilization in the gas chromatograph, this criterion unfortunately cannot be met. We are reminded of some work conducted in 1952 in which separation was performed by liquid upon silica gel and the rancid components of reverted soybean oil were separated from the painty components (8). It would be hoped with the modern developments of high-pressure liquid chromatography (HPLC), and with the developments of new phases, that it will be possible to separate unchanged the sensitive mixture of odor components. The current problem of this approach is that of interfacing the high-pressure liquid chromatograph to the mass spectrometer. It is a problem of separating the volatile odor materials from a frequently equally volatile developing solvent. At the present time the advances in our scientific knowledge are being limited by needed advances in technique.

The disposition of U.S. soybean meal as shown in Figure 4 differs markedly from that of the oil. From this chart and the relative dollar value of oil and meal components of the soybean, it may be inferred that the soybean crop is raised primarily for its protein content and for its outlet in the animal feeding economy. It is of interest that the nonfeed uses of soybean meal, which include 2% for human food, amount to only 5% of the total disposition. There is a need worldwide by protein-starved nations for just this kind of high-quality, low-cost protein which soybeans provide. This statement then begs the question of why there is not more volume of edible soy products, particularly in the light of many commercially available products in the United States which contain soybean meal or constitute food-grade soybean meal for mixing purposes.

Some of the critical factors that restrict the use of soy meal in foods are legalistic--about which research has little to say. There are, however, problems of functional properties and of form. There are problems of making available maximal nutritional values by processing treatments. For example, a soybean meal that is designed for minimal denaturation and maximum solubility may have anti-metabolites, or its amino acids may be less available

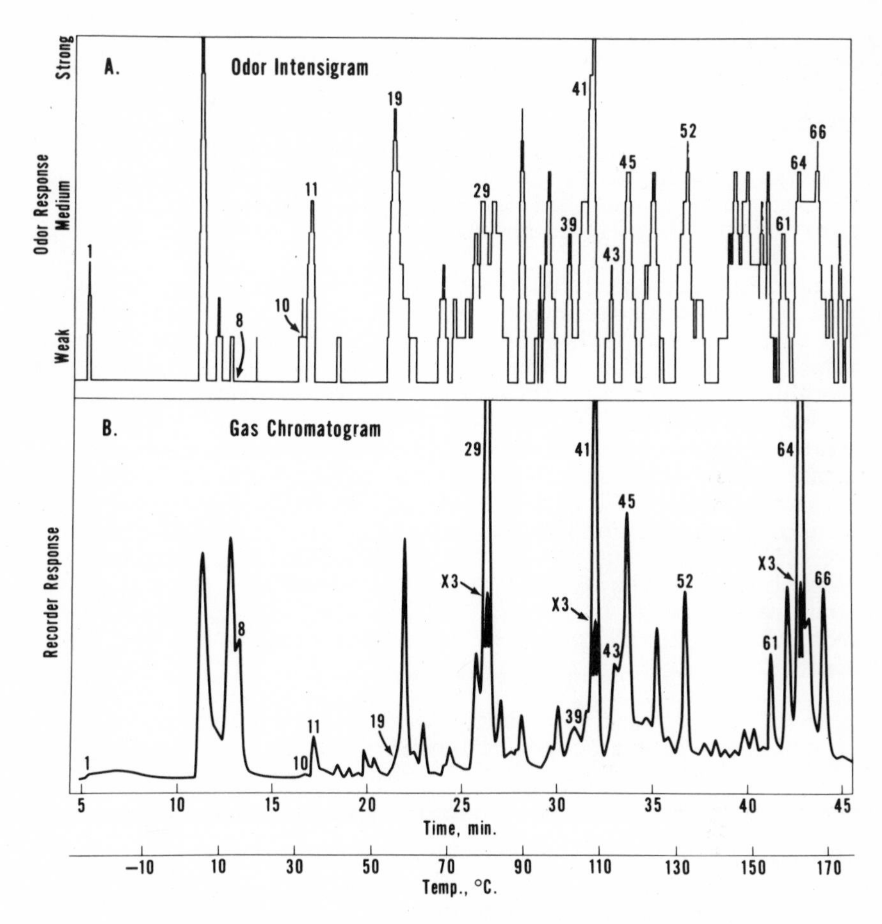

Figure 3. (Top) odor intensiogram or odor intensity of peaks corresponding to the chromatogram of volatiles (bottom) collected for 10 min immediately after soybean oil reached 193°C

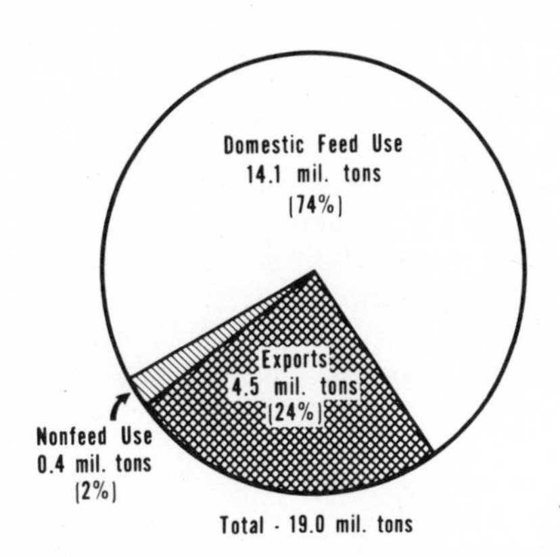

Figure 4. Disposition of U.S. soybean meal—1976

than if it were heat treated. However, this very process of toasting will reduce the solubility characteristics; thus, there must be a trade-off or compromise between solubility and its nutritional value.

A problem sometimes encountered in soybean protein use is that of gas production or flatus. Research has established that the galactoside type of tri- and tetrasaccharides, which are present in soybean meal to the extent of 10%, are indicted in flatus (9).

Certainly the most important factor that restricts the widespread food use of soybean meal at the present time is its flavor. It is well known that this is a critical factor and that starving populations will not eat nutritious materials if the flavor is not acceptable to them. In some bland products where it is used as an extender, soybean meal is limited by flavor to ca. 30% addition. Therefore, it is in this area of reduction of flavor that the greatest effort of research should be concentrated.

Given in Table II is a listing of flavor scores recorded by the taste panel for commercial products, for the concentrates and for the isolates. A score of 10 is indication of a bland product. In most instances the alcohol treatment to produce the concentrate or the precipitation to produce the isolate gives an improved flavor, as well as improved flatus characteristics, to the product (10).

Table II
Odor and Flavor Scores of Hexane-Defatted Soy Flours

Sample	Odor scores	Flavor scores
Commercial flours, A-G	5.8-7.5	4.2-6.6
Raw flour	5.8	4.1
Azeotrope-extracted		
Hexane:methanol	6.2	6.1
Hexane:ethanol	7.9	7.2
Hexane:2-propanol	5.9	5.1

The flavor of soybean meal comes from two general sources: first, it is indigenous to the bean itself and, secondly, it may be formed upon storage. As shown in Table III, the flavor intensity values for the beany characteristic of maturing soybeans remained relatively constant during the maturation process (11). However, the bitter principle increased rather sharply as the soybeans matured. Many of the compounds contributing to the green flavors of raw soybeans are known, and some of the principal ones are listed in Table IV (12).

Table III
Flavor Intensity Values (FIV)
of Maturing Soybeans

Days After Flowering	Beany		Bitter	
	%[a]	FIV[b]	%[a]	FIV[b]
28	100	2.0	25	0.42
34	90	2.3	50	0.90
49	100	2.5	82	1.4
56	100	2.2	86	1.9
63	100	2.7	100	2.2
66	92	2.5	100	2.1

[a] Percentage of panelists giving a positive response.
[b] Based on score 1 = weak to 3 = strong.

Table IV
Key Compounds Contributing to Green Flavors of
Raw Soybeans and Peas

Compound	Flavor description	Threshold (ppm in oil)	
		Odor	Taste
n-Hexanal	Green grassy	0.32	0.15
3-*cis*-Hexenal	Green beany	0.11	0.11
n-Pentylfuran	Beany	2	1-10
Ethyl vinyl ketone	Green beany	5 (milk)	

The flavors mentioned can also arise from deteriorative reactions either in the processing or storage of the soybean meal. These reactions may be of an enzyme-catalyzed nature as for example by lipoxygenase or they may come about by autoxidation in a manner not greatly dissimilar to that in the oil (13). In any case, the hydroperoxides thus formed may undergo decomposition to form the keto or hydroxy type of functional groups as postulated in Figure 5, or even the epoxide type of compound as is shown in Figure 6.

Some of the major volatile compounds derived from linoleic acid by lipoxygenase and analyzed in headspace are shown in Table V. These chemical compounds are readily rationalized by the oxidative mechanisms of breakdown of hydroperoxides of linoleic acid (14).

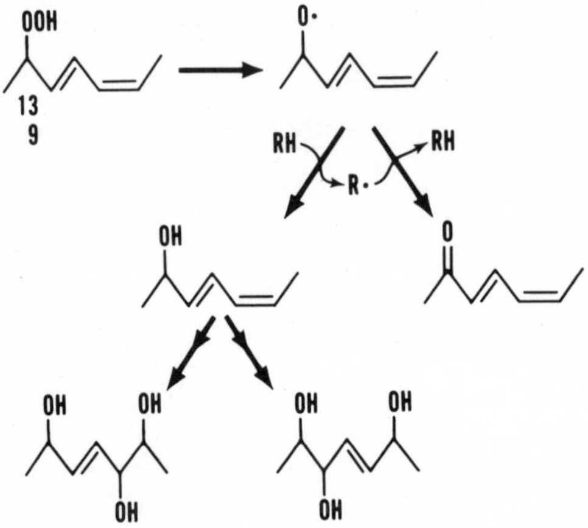

OOH
13
9
O·
RH
R·
RH
OH
O
OH OH
OH
OH OH
OH

Figure 5. Postulated decomposition of unsaturated fatty acid hydroperoxides

1) ROOH + R₂C=CR₂ → ROH + R₂C—CR₂ (epoxide, O)

2) OOH → O· → epoxide radical → ·OH → OH epoxide

Fe+++

+ → HOH → OH epoxide

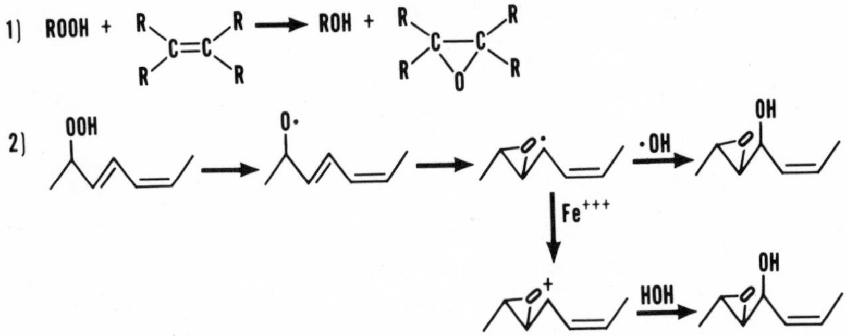

Figure 6. Postulated formation of epoxides from fatty acid hydroperoxides

Table V
Major Volatile Compounds Generated from Linoleic Acid Hydroperoxides (LOHP) by Pea and Soybean Lipoxygenases[a]

Compound	Quantity in headspace, GLC peak height[b]	
	Pea	Soybean
n-Butanal	+	+++
n-Pentanal	+++	+++
n-Hexanal	+++	+++
n-Heptanal	++	+
n-Hept-*trans*-2-enal	+++	+++
2-n-Pentyl furan	+	+++

[a] See Ref. 15.
[b] + = 5-10 mm; ++ = 11-50 mm; +++ = >50 mm. GLC = gas liquid chromatography.

For soybean meal as compared to other vegetable meals, the linolenic acid is an unusual component of the residual fat. When the linoleate hydroperoxides were tasted by our taste panel, rancidity was listed as one of the most frequent responses. By contrast, when linolenic acid was so treated, rancidity was a less prominent response. The grassy-beany flavors, which may be assumed to include paintiness as well, were the most important response in the linolenate hydroperoxides (13).

Although it has not been established, it would appear that oxidized soybean phosphatidylcholine contributes to the bitterness of the soybean meal. This has been studied by oxidizing purified phosphatidylcholine and submitting to the taste panel at various levels. It has also been indicated by fractionation studies in which the bitter principle from hexane-extracted soy flour was found to be concentrated in the purified soybean phosphatidylcholine fraction (15). While all of this basic type of research is being conducted on the flavor of the soybean meal, industry has been implementing some of the research results generated 30 years ago. At that time, it was established that ethanol extraction of the flakes had a particular merit in removing bitter principles from soybean meal (16). Illustrated in Table II, the hexane-ethanol azeotropic solvent was capable of removing the residual lipids and the bitter flavor. It was noted that washing with diethyl ether by itself did not remove bitter flavor, but not until the hot ethanol treatment was given were the bound lipids removed which contained the highly flavored principles. As shown in the Table, many of the commercial flours had odor and flavor scores not greatly different from raw soybean flour. Upon extraction by the azeotropic alcohol mixtures the flavor was improved, but of those investigated the hexane-ethanol seemed to have a peculiarly beneficial effect in improving the flavor. In fact, the combination of toasting and azeotropic extraction in the laboratory appears to have raised the flavor score for odor and flavor to that equivalent to wheat flour. The flavor intensities of the grassy-beany and the bitter flavor were reduced. It has been inferred that alcohol treatment may be responsible for some of the improved flavor of soybean meal products now being produced.

In summary, then, the flavor stability of soybean oil which was regarded as the number one problem of the soybean industry three decades ago has now, at least in part, been solved; but the flavor of the meal is today held as the major deterrent to the increase of soybean protein in human food products. Since residual lipids of the soybean flakes constitute the precursors for odors and flavors, knowledge of deteriorative factors of fats is basic to the understanding of flavor development in both soybean meal and oil products. Highly sophisticated gas chromatographic-mass spectrometric systems and HPLC are being used for analysis of odors of oil and meal products, and these techniques are particularly effective where the human nose is included in the computer loop. Meanwhile current applications of past research on the processing of soybean oil and meal appear to have made a significant contribution in solving the respective problems for the food industry.

Literature Cited

1. Moser, Helen A., Jaeger, Carol, Cowan, J. C., and Dutton, H. J. J. Am. Oil Chem. Soc. (1947) 24:291.
2. Dutton, H. J., Schwab, A. W., Moser, Helen A., and Cowan, J. C. J. Am. Oil Chem. Soc. (1948) 25:385.
3. Dutton, H. J., Schwab, A. W., Moser, Helen A., and Cowan, J. C. J. Am. Oil Chem. Soc. (1949) 26:441.
4. Dutton, H. J., Lancaster, Catherine R., Evans, C. D., and Cowan, J. C. J. Am. Oil Chem. Soc. (1951) 28:115.
5. Koritala, S., and Dutton, H. J. J. Am. Oil Chem. Soc. (1966) 43:86.
6. Cowan, J. C., Evans, C. D., Moser, Helen A., List, G. R., and Koritala, S. J. Am. Oil Chem. Soc. (1970) 47:470.
7. Selke, E., Rohwedder, W. K., and Dutton, H. J. J. Am. Oil Chem. Soc. (1972) 49:636.
8. Kawahara, F. K., and Dutton, H. J. J. Am. Oil Chem. Soc. (1952) 29:372.
9. Steggerda, F. R., Richards, E. A., and Rackis, J. J. Proc. Soc. Exp. Biol. Med. (1966) 121(4):1235.
10. Rackis, J. J., Eldridge, A. C., Kalbrener, J. E., Honig, D. H., and Sessa, D. J. AICHE Symp. Ser. (1973) 69(132):5.
11. Rackis, J. J., Honig, D. H., Sessa, D. J., and Moser, H. A. Cereal Chem. (1972) 49:586-597.

12. Sessa, D. J., and Rackis, J. J. J. Am. Oil Chem. Soc. (1977) 54:468-473.
13. Kalbrener, J. E., Warner, K. A., and Eldridge, A. C. Cereal Chem. (1974) 51:406-416.
14. Leu, K. Lebensm.-Wiss. U. Technol. (1974) 7:88.
15. Sessa, D. J., Warner, K. A., and Rackis, J. J. J. Agric. Food Chem. (1976) 24(1):16-21.
16. Belter, P. A., Beckel, A. C., and Smith, A. K. Ind. Eng. Chem. (1944) 36:799.

RECEIVED April 10, 1978

8

Flavors from Lipids by Microbiological Action

L. WENDELL HAYMON
Microlife Technics, P.O. Box 3917, Sarasota, FL 33578

JAMES C. ACTON
Clemson University, Clemson, SC 29631

Flavors from lipids are ubiquitous throughout the foods presently consumed in the world. In this symposium, we have learned of flavors generated by frying, smoking, and a wide range of chemical reactions.

This part of the symposium will review the role of microorganisms in the development of flavors from lipids, particularly the fermented sausage industry.

The complex nature of microbiological reactions is shown in "Figure 1". It is often difficult to separate the reactions caused by the action of microbes. Many flavor components come from the oxidative reactions which are initiated following the hydrolytic activity of microbes. "Figure 1" shows the hydrolytic action on lipid material to free fatty acids by microbiological action. The liberation of these free fatty acids is accompanied by an increase in total carbonyls and peroxide values.

Microorganisms are responsible for many types of flavors in meat products, particularly fermented sausage products and cheese products. The flavor compounds which are produced from the animal fat, are directly responsible for the varied taste of semi-dry and dry sausages. The lipases and other enzymes that cause hydrolysis of lipids in animal fat are primarily responsible for the generation of flavors from lipids in meat products. The enzymes and lipases are made by the microbe in the meat product.

Cheeses are the fermentation products of the dairy industry which also are high in lipid content and therefore are candidates for lipase activity of milk fats. The fatty acid distribution of cheeses can be changed by the selection of the microorganism used as a starter culture, or by the indigenous flora of milk.

Blood (1975) reported on the growth of lactic acid bacteria in fish products. The actions of the microorganisms were lipolytic and proteolytic. The final reactions and conclusions of the role of lactic acid bacteria in marinade of fish is not well understood at this time.

0-8412-0418-7/78/47-075-094$10.00/0

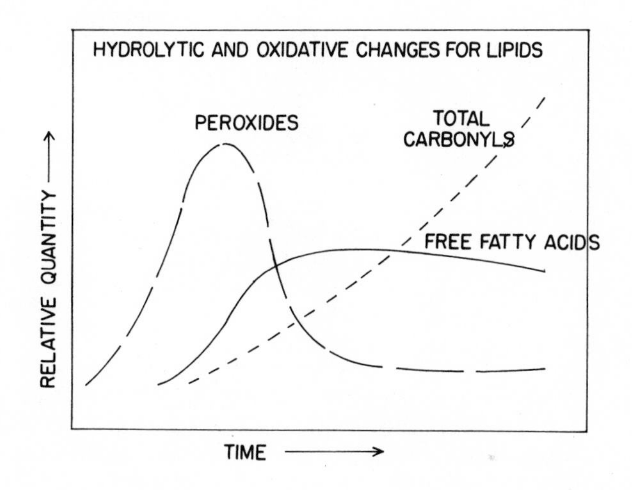

Figure 1. General sequence of lipid fraction changes attributed to hydrolysis and oxidative activities

Yogurt is a special class of dairy product since it is now being sold in greater volumes than ever as a frozen dessert item. Yogurt is a fermented milk product which is usually fermented by two organisms, L. bulgaricus and Streptococcus thermophilus. The role of the organisms is to lower the pH of the product and to flavor the product. The by-products of the lactic acid fermentation (see "Figure 2") influence the taste of yogurt, and off-flavors and odors are caused by faulty fermentation procedures. Very little research effort has been devoted to the lipid changes in yogurt.

There are flavor changes in bread which are due to the microbiological action of yeast. Within this decade, a lactic acid bacteria (L. san francisco) was shown to contribute to the flavor and fermentation of sour dough bread. Wood (1975) has described the action of the lactic acid bacteria in bread making and prompted the analogy of the sour rye bread of Scotland and sour dough bread of San Francisco. The action of the lactic acid bacteria appears to be through direct utilization of maltose. However, these organisms are known to possess active lipase enzyme systems and their action on bread lipids may be quite beneficial.

Microbiological Action on Lipids

"Figure 3" shows the typical microbiological lipase action on the mono, di or triglyceride. The action of lipase is usually at 15°-40°C, and results in an increase in free fatty acids, and either mono- or diglycerides depending on the specificity of the lipase. The same hydrolytic action of lipase can be accomplished at higher temperatures in an acidic media. Lactic acid bacteria provide such a media for the fermented foods described earlier.

As shown in "Figure 1", the accumulation of free fatty acids in a food caused by the action of microbioligical lipase can be accompanied by an increase in TBA value (thio-barbituric acid value), peroxide value, or total carbonyls. These indicators show that oxidative reactions were initiated, often due to the action of lipase. Also, these reactions can be initiated by hydrogen peroxide. Lactobacilli can be producers of hydrogen peroxide (Tjakerg, 1969). Oxidative reactions are usually associated with rancid off flavors; however, some oxidation of lipids is necessary to get a balance of flavors or to create a flavor for a product.

Biochemical Reaction of Bacteria

Lactic Acid Bacteria. The metabolism of lactic acid bacteria is well established. They metabolize simple sugars to lactic acid (homo fermentative) or lactic acid, acetic acid, and carbon dioxide (heterofermentative). Bergy's 8th Edition (1974) lists the sugars metabolized by lactic acid bacteria. There is no system to identify the specific lipase system of bacteria. Some

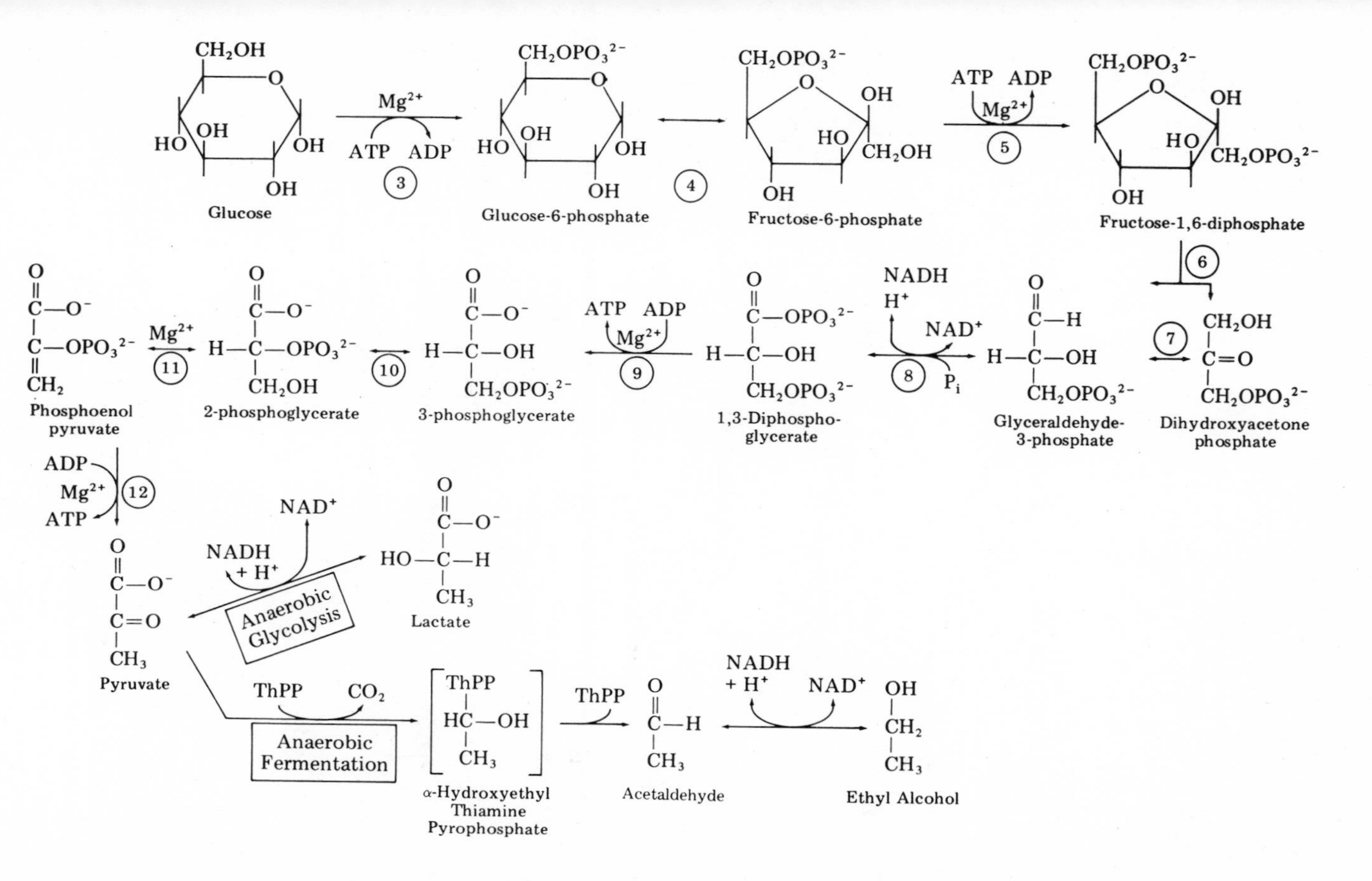

Figure 2. Structures, enzymes, and cofactors for glycolysis and fermentation

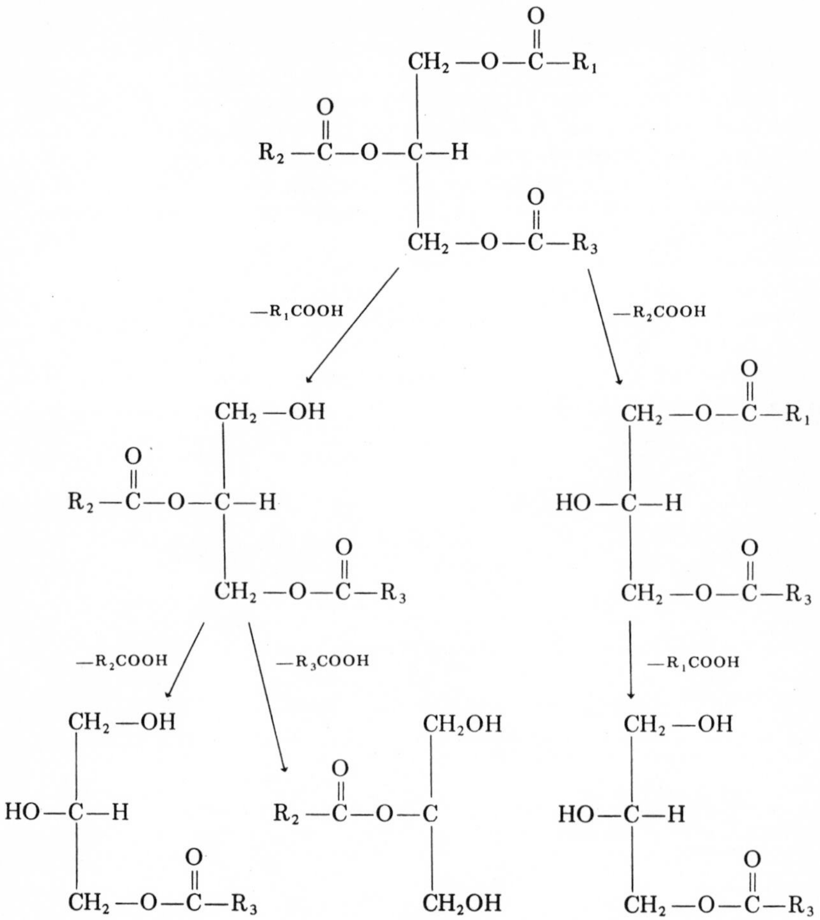

$$\begin{array}{l} CH_2—O—\overset{O}{\overset{\|}{C}}—R_1 \\ R_2—\overset{O}{\overset{\|}{C}}—O—C—H \\ CH_2—O—\overset{O}{\overset{\|}{C}}—R_3 \end{array}$$

$—R_1COOH$ ↙ ↘ $—R_2COOH$

$$\begin{array}{l} CH_2—OH \\ R_2—\overset{O}{\overset{\|}{C}}—O—C—H \\ CH_2—O—\overset{O}{\overset{\|}{C}}—R_3 \end{array} \qquad \begin{array}{l} CH_2—O—\overset{O}{\overset{\|}{C}}—R_1 \\ HO—C—H \\ CH_2—O—\overset{O}{\overset{\|}{C}}—R_3 \end{array}$$

$—R_2COOH$ ↙ ↘ $—R_3COOH$ ↓ $—R_1COOH$

$$\begin{array}{l} CH_2—OH \\ HO—C—H \\ CH_2—O—\overset{O}{\overset{\|}{C}}—R_3 \end{array} \qquad \begin{array}{l} CH_2OH \\ R_2—\overset{O}{\overset{\|}{C}}—O—C \\ CH_2OH \end{array} \qquad \begin{array}{l} CH_2—OH \\ HO—C—H \\ CH_2—O—\overset{O}{\overset{\|}{C}}—R_3 \end{array}$$

Complete hydrolysis will yield glycerol and free fatty acids.

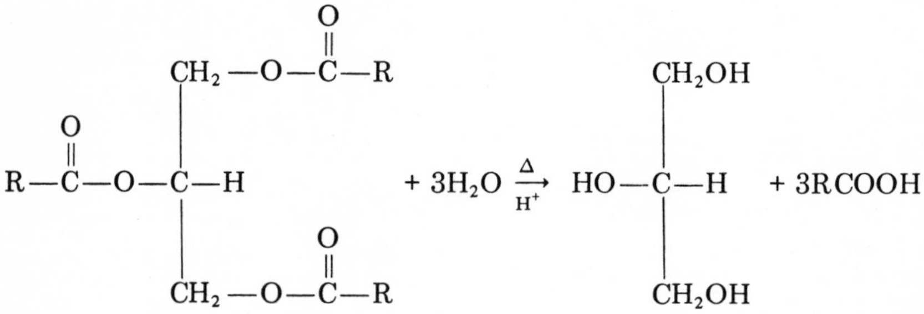

$$\begin{array}{l} CH_2—O—\overset{O}{\overset{\|}{C}}—R \\ R—\overset{O}{\overset{\|}{C}}—O—C—H \\ CH_2—O—\overset{O}{\overset{\|}{C}}—R \end{array} + 3H_2O \xrightarrow[H^+]{\Delta} \begin{array}{l} CH_2OH \\ HO—C—H \\ CH_2OH \end{array} + 3RCOOH$$

Figure 3. Actions of lipase

lactic acid bacteria are active lipase producers, while others of the same classification do not produce significant lipase activity.

Alford, Smith and Lilly (1971) have shown the hydrolytic and oxidative changes due to Micrococci, Pseudomonas, and Staphylococci. They showed that the hydrolytic activity of microorganisms is dependent on (1) the fat used as a substitute, (2) temperature of incubation, (3) composition of the growth media, and (4) oxygen availability.

Table I from Alford & Pierce (1961) shows the effect of temperature on the type of fatty acids released by microbial lipases from coconut oil. Lipases of bacteria isolated from rancid butter (by Cooke 1973) showed that Pseudomonas, S. aureus, Micrococcus saprophyticus, Micrococcus spp., and a pancreatic lipase produced varying amounts of C_{12}, C_{14}, C_{16}, C_{18}, $C_{18:1}$ fatty acids from milk fat at 35, 30, and 22°C. In contrast to earlier findings, Micrococcus produced a large increase of oleic acid at high temperatures.

A high carbohydrate substrate will inhibit or reduce lipase production (Nashif & Nelson, 1953; Alford & Elliott, 1960) and the protein, peptides or amino acids used as sources of nitrogen are important consideration (Lawrence et al., 1967; Alford & Pierce, 1963).

TABLE I

Influence Of Temperature On Type Of Fatty Acids Released By Microbial Lipases From Coconut Oil (After Alford & Pierce, 1961)

Lipase From	% Of Total Free Fatty Acids* Released As			
	Lauric Acid At		Oleic Acid At	
	35°	-7°	35°	-7°
Pseudomonas Fragi	46	29	5	25
Geotrichum Candidum	39	3	14	52
Penicillium Roquefortii	46	35	4	24

*Conditions of incubation of enzyme-coconut oil mixtures and % of other free fatty acids released are given in the original paper.

Vigorous aeration decreases lipase production or at least its accumulation, yet growth in media with high surface/volume ratios such as sausages or with slow agitation is stimulatory (Nashif & Nelson, 1953; Alford and Elliott, 1960; Alford & Smith, 1965; Lawrence et al., 1967).

Lipase Specificity

Table II shows the determination of specificity of microbial lipases by their action on triglycerides of known composition (Alford et al., 1964). There appears to be good evidence of specificity related to the position of attachment of a fatty acid to the triglyceride molecule (Alford et al., 1964; Mencer & Alford, 1967) and to the structure of the fatty acid being hydrolyzed (Jensen et al., 1965). Other lipases are very general with no specificity. The *Geotrichum candidum* is specific for oleic acid regardless of position, and *Ps. fragi* lipase is specific for the 1 - position regardless of the fatty acid, and *S. aureus* lipase exhibits neither position nor fatty acid specificity.

TABLE II

Determination Of Specificity Of Microbial Lipases
By Their Action On Triglycerides Of Known Composition
(After Alford, Pierce & Suggs, 1964)

	% Composition Of Triglyceride			
Lipase From	2-Oleoyl Distearin		2-Stearolyl Diolein	
	Oleic Acid	Stearic Acid	Oleic Acid	Stearic Acid
Geotrichum Candidum	98*	2	99	1
Pseudomonas Fragi	2	28	98	2
Staphylococcus Aureus	25	75	63	37

* % Of total free fatty acids released from enzyme-triglyceride mixture incubated at 35° from 1-3 H.

Smith & Alford (1966) have shown that the production and activity of some lipases are sensitive to end-product accumulation inhibition. Their studies were performed with *Ps. fragi*, and the end-product inhibition was eliminated by the addition of bovine serum.

Oxidative Reactions

Alford and Smith (1971) have shown the action of microorganisms on the peroxides of rancid lard, shown in "Figure 4". The effect varied from 18% decomposition by Ps. ovalis and Streptomyces sps. to 100% decomposition of the peroxides by G. candidum and Aspergillus flavus. Fifteen cultures of the 29 studies showed a 50% reduction in the peroxide value. The action of microorganisms on the mono carbonyls was shown by Alford and Smith (1971). "Figure 5" and "Figure 6" show that Ps. fragi, Ps. ovalis, and G. candidum destroyed 100% of the 2, 4 dienal fraction while Asp. flavus and M. freudenreichii increased the dienal content 4 to 7 fold. Fifteen cultures of the 29 studied decreased the dienal concentration. Only 5 cultures of the 29 increased the dienal concentration by a factor of 2.

"Figure 7" shows the effect of microorganisms on peroxides in fresh lard. Ten of the 28 microorganisms evaluated destroyed the small amount of peroxide present in fresh lard, but 14 had no effect. Five strains of Streptomyces increased the peroxide concentration about 3 fold, Ps. ovalis increased the concentration by 8 fold, and M. freudenreichii increased the peroxides about 14 fold.

Alford & Smith (1971) reported that M. freudenreichii produced a large increase in the concentration of 2, 4 dienals and 2 enals. Most of the microorganisms had little effect on the alkanal fraction. Ps. fragi, G. candidum, and C. lipolytica increased the concentration of the alkanals as well as producing methyl ketones, a fraction not present in fresh lard. The ability of Streptomyces spp., Ps. ovalis, and M. freudenreichii to form peroxides suggests that lipoxidase like activity is present. It was reported that microorganisms produce lipoxidase (Mulkerjee, 1951; Fukuba, 1953; Shimahara, 1966), but Tapel (1963) stated there is no evidence for a microbial lipoxidase.

The precursors for methyl ketone production by the studied bacteria are unknown, but these organisms are strongly lipolytic and fungi are known to produce methyl ketones by B-oxidation and decarboxylation of lipase liberated fatty acids (Hanke, 1966).

Sausage Products

The type of fat used in the preparation of dry sausage will influence flavor characteristics, particularly as the fat shifts from all beef formulations to all pork formulations. The distinctive flavors of dry sausages are due in part to the hydrolytic and oxidative changes that occur in the lipid fraction during ripening or drying.

Lipase Activity. The hydrolytic changes in fats are due primarily to the action of bacteria which produce lipases. These microbial lipases act to free fatty acids and glycerol. In dry

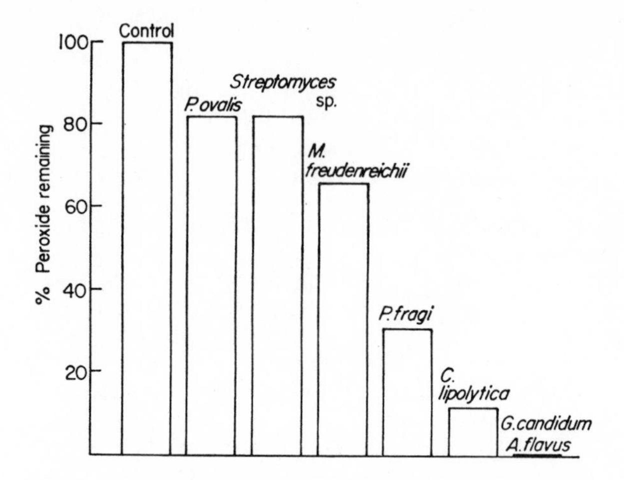

Figure 4. The effect of micro-organisms on the peroxides of rancid lard. The initial peroxide values ranged 74.3–97.4 meq/kg of fat (mean, 82.9 meq).

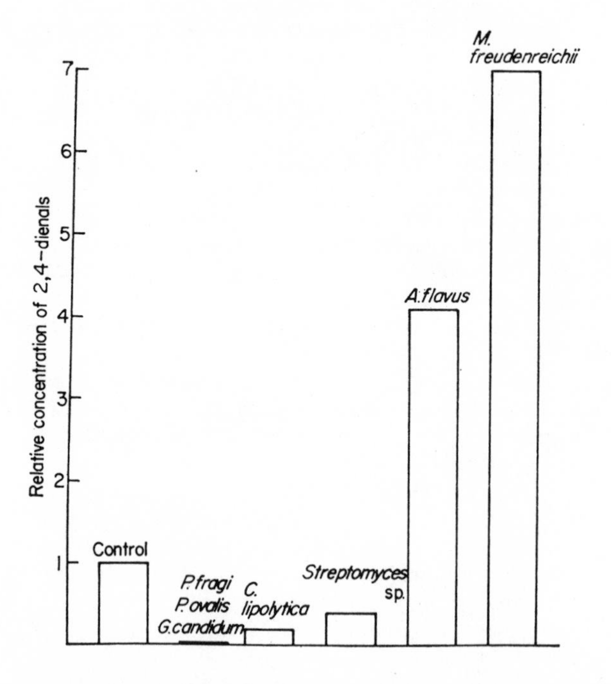

Figure 5. The effect of micro-organisms on the 2,4-dienals of rancid lard. The 2,4-dienal concentration of the controls ranged 1.5–3.4 μmoles/10^4 μmoles fat (mean, 2.0 μmoles).

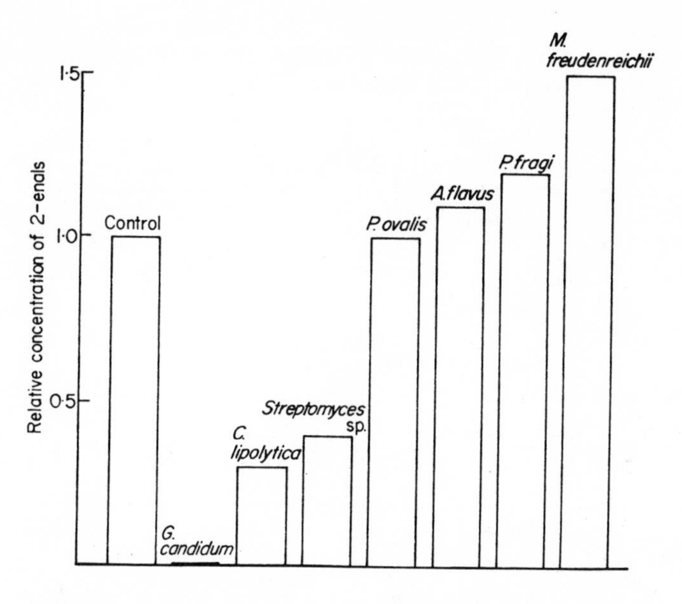

Figure 6. The effect of micro-organisms on the 2-enals of rancid fat. The 2-enal concentration of the controls ranged 3.6–6.7 μmoles/10⁴ μmoles fat (mean, 5.0 μmoles).

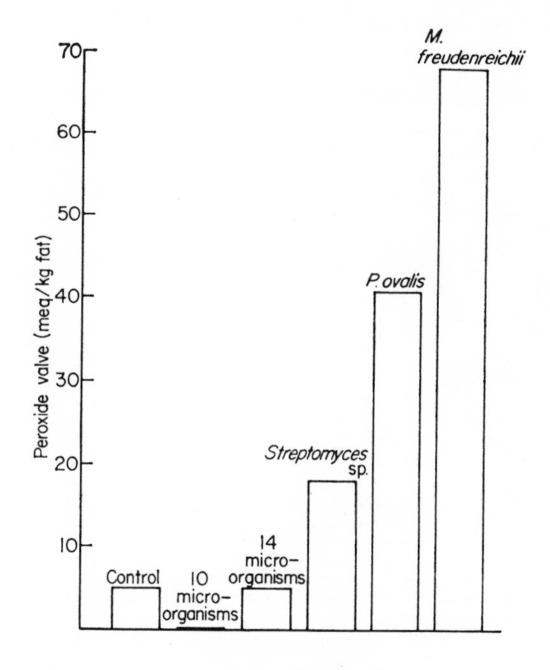

Figure 7. The effect of micro-organisms on peroxides in fresh lard

sausage, where no heat is applied to the product, hydrolysis may occur through muscle and adipose tissue lipases (Wallach, 1968). The particular flavor developed by the sausage via lipase activity depends on the composition of the fat as seen from the work of Alford and Smith (1971).

The genus _Micrococoiccus_ is generally accepted as being the predominant group of microbes responsible for hydrolysis of fats in dry sausage (Table III) (Cantoni _et al._, 1967), but recent studies show that some species of _Lactobacilli_ produce very active lipases at 20°C and higher (Stoychev _et al._, 1972a, 1972b; Covetti, 1965).

TABLE III

Action of Micrococci on Pork Fat
(from Cantoni _et al._)

	Culture	
	D10	C13
C_4 - C_{20}, G FA/100 G FAT[A]	23.1	45.2
(FA most easily released: Oleic, Myristic, Palmitoleic, Linoleic)		
Volatile FA, MG/100 G FAT[B]	87	50
(principal VFA: Propionic, Acetic)		
Carbonyls, μM/l of Culture[C]	660	667
(principal carbonyls: Propionaldehyde, Isovaleraldehyde)		

[A]After 28 days of culture; FA - Fatty Acid.

[B]After 28 days of culture; VFA - Volatile Fatty Acid.

[C]After 24 days of culture.

Mihalyi and Kormendy (1967) reported an increase in free fatty acid values in the inner and outer zones of a Hungarian dry salami aged for 100 days (Table IV). The outer zone showed a higher level of fatty acids than the inner zone. Since there is mold accumulation on Hungarian salami, the increase was attributed to this mold growth.

TABLE IV

Free Fatty Acid Values (mg KOH/g fat)
of Hungarian Dry Salami During Ripening*

Days Of Ripening	Sausage Portion	
	Inner Zone	Outer Zone
10	6.07	2.88
40	11.28	12.40
70	14.81	18.29
100	17.68	20.62

*Adapted from Mihalyi and Kormendy (1967).

Lu and Townsend (1973) also showed increases in free fatty acid values during the drying cycle and correlated their results with parallel peroxide values (Table V).

Demeyer *et al.* (1974) reported that linoleic acid was liberated at a faster rate than all of the other acids in a pork dry sausage ("Figure 6"). Brockerhoff (1966) has shown that pork triglycerides have about 60% of the stearic acid located in position 1, palmitic acid ($\sim$ 60-80%) at position 2, and octadecenoic acid ($\sim$ 50-60%) are at position 3 of the triglyceride molecule. Demeyer *et al.* (1974) found the rate of hydrolysis to free fatty acids decreased in the following order: linoleic > oleic > stearic > palmitic. The lipase generally had a specificity for position 3.

Cerise *et al.* (1973) reported that oleic acid was the principal free fatty acid found in the lipid fraction of Italian pork salami. Dobbertin *et al.* (1975) reported yeasts, pseudomonads, and enterococci which exhibited high lipase activity, but the lactobacilli had little or no lipase activity. Dobbertin concluded that lipase activity was independent of total bacterial count. However, Coretti (1965) found that some lactobacilli appear to form lipase activity adaptively in later generations. Fryer et al. (1967) and Oterholm (1968) have reported lipolytic activity for lactobacilli of tributyrin, but not triglycerides.

TABLE V

Free Fatty Acid Values (mg KOH/g fat) and Peroxide Values of a Fermented Dry Salami*

Days Of Ripening	Free Fatty Acid Value	Peroxide Value
0	3.08	10.3
14	6.39	17.4
21	6.90	12.9
28	6.56	16.9
35	11.66	20.6

*Calculated from data for control sausages in study of Lu and Townsend (1973).

The Pseudomonas, Micrococci, and Staphylococci are active lipase organisms. Debevere et al. (1976) from Belgium have reported the isolation of a Micrococcus species from a starter culture used for producing dry sausages, which has a strong lipolytic activity on pork fat. This species caused the release of fatty acids from pork fat in a nonspecific way but in about the same proportions that occur in pork fat. Both saturated and unsaturated carbonyls were formed from the long chain fatty acids. The unsaturated carbonyls disappeared faster than the saturated carbonyls, apparently due to oxidation. The breakdown of the saturated carbonyls appeared to be due to the action of the bacteria, perhaps by producing enzymes capable of hydrolyzing these compounds. The bacteria may play an important role in the development of flavor by releasing and hydrolyzing the carbonyl compounds, and producing short chain volatile compounds that contribute to flavor.

Oxidative Changes in Sausages. Cerise reported two distinct phases of lipid changes. In the early ripening phase of fermentation, lipase activity occurred to yield free fatty acids, which were in turn oxidized by peroxides to carbonyl compounds in the drying phase. Free fatty acids decreased as the amount of carbonyl compounds increased. Cerise (1972) showed that peroxide values decrease rapidly after the fermentation phase ("Figure 8"). Peroxide values increased dramatically between the second and fourth days at 21°C and remained high before decreasing to below initial levels at 15 days of ripening.

Nurmi (1966) reported nearly equivalent amounts of peroxide formation at 3 days of sausage ripening with Micrococci and/or lactobacilli. After the 3 day fermentation period, sausages containing lactobacilli continued to show peroxide at higher levels than initially observed. Both micrococci and lactobacilli can be strong producers of peroxide. While micrococci are catalase-positive, lactobacilli are catalase-negative. Nurmi (1966) pointed out that faulty product flavor and color may result when catalase-negative lactobacilli are used as starters for sausage ripening.

Hydrogen peroxide, which is formed by the pediococci, and lactobacilli, is an available and willing reactant in the oxidation of free fatty acids generated by lipolysis, especially unsaturated fatty acids (Tjaberg, 1969). Also, fatty acid oxidation may result from autooxidation mechanism. Carbonyl compounds are formed as a direct result of unsaturated fatty acid decomposition. The evaluation of total carbonyl compounds for a dry sausage made primarily with pork fat is shown from Demeyer et al. (1974).

Halvarson (1973) qualitatively identified some 22 volatile carbonyl compounds from Swedish fermented sausage. The predominant substances were ethanal, propanal, propanone, and 2-methyl and 3-methyl butanal in concentrations of 0.6 to 3.6 mg. per kg. of sausage. All the straight chain alkanals were detected up to octanal, with no butanal detected, as well as methyl ketones, 2-alkenals, and 2, 4-alkadienals. Langer et al. (1970) reported qualitative identification of 29 carbonyl compounds during the ripening of a dry salami. Langer et al. (1970) and Halvarson (1973) considered the lower molecular weight carbonyls (probably from carbohydrate fermentation) to possess minimal values for characteristic sausage aroma, and both authors state that the main aroma producers are the higher molecular weight carbonyls. In particular, unsaturated carbonyls such as 2-alkenals and the 2, 4-alkadienals, are potent flavor compounds typically present in oxidized pork fat, and present to a limited extent in oxidized beef fat (Hornstein and Crowe, 1960; 1963).

Cheese

Cheese is a fermented dairy product which has a high fat content. The lipid changes in cheese must help determine the final flavor of this highly flavored product. Sometimes, flavors

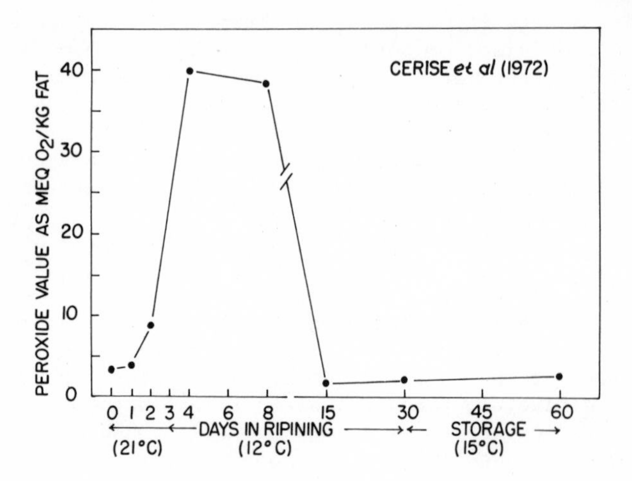

Figure 8. Peroxide values during sausage ripening (0 to 3 days at 21°C, then stored at 12°C) (adapted from Ref. 13)

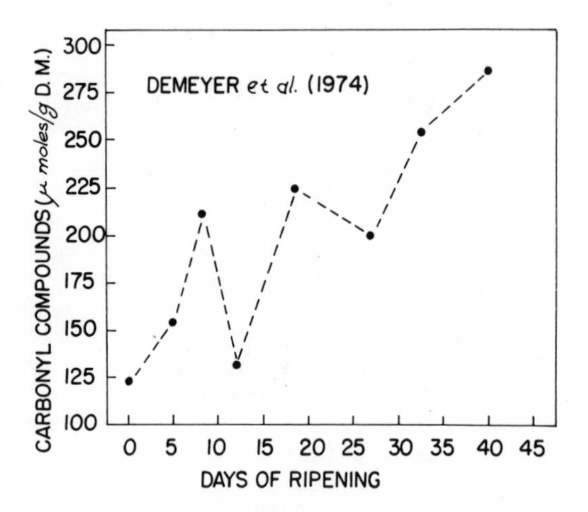

Figure 9. Changes in total carbonyl content (as 2,4-dinitrophenylhydrazones) during dry sausage ripening (adapted from Ref. 20)

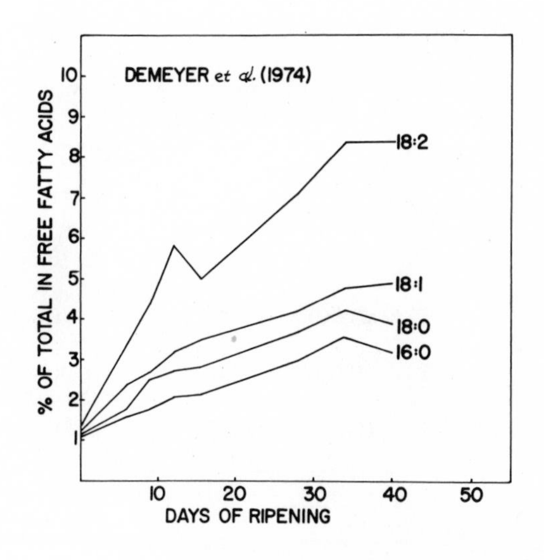

Journal of Food Science

Figure 10. Percent of total palmitic (16:0), stearic (18:0), oleic (18:1) and linoleic (18:2) acid present in the free fatty acid fraction (20)

of cheeses are described as soapy, bitter, metallic, barn-yard (butyric), or rancid. These flavors are caused by the accumulation of free fatty acids. The short chain fatty acids have low taste thresholds and need only be present in minute concentrations to produce flavors. Deeth and Firzgerald (1976) have reviewed the literature thoroughly for lipolysis in milk-products.

TABLE VI

Volatile Fatty Acids Reported In Dry Sausages

Acid	Deketelaere et al. (1974)	Halvarson (1973)	
		Nonsmoked	Smoked
		mg/g sausage	
Formate	-	0.25	0.42
Acetate	2.4 mM/100 g D.M.	0.70	1.2
Propionate	11.7 μM/100 g D.M.	0.004	0.03
n-Butyrate	14.5 μM/100 g D.M.	0.004	0.007

The most common source of lipase other than milk lipase is from psychrotrophic bacteria, those which grow at refrigeration temperatures. Deeth states that "when the count of these lipolytic bacteria exceed one million per milliliter they cause rancid flavors." However, cheeses such as Romano, the Parmesan and blue vein types depend for their distinctive tastes upon relatively high levels of particular fatty acids which are produced by rennet or microbial and fungal lipases during maturation.

The cheddar cheese flavor is reported by Deeth (1976) as a balance between fatty acids produced in low amounts during normal aging and other flavor constituents.

Peterson and Johnson (1949) isolated 12 of 54 lactobacilli which possessed intracellular lipase active between pH 5 and 6, and were capable of butterfat hydrolysis. L. casei (four strains) was particularly active and liberated butyric, caproic, caprylic, and capric acids from butterfat.

Morris and Jezeski (1953) characterized the lipase system of Pencillium roqueforti. Lawrence and Hawke (1968) found that the fatty acids liberated from P. roqueforti gave oxidation products dependent on the fatty acid concentration, chain length of fatty acids, and pH of the system. The studies were conducted by oxygen uptake.

Lubert and Frazier (1955) studied cultures of film yeasts and of micrococci from brick cheese and cheese brines. The micrococci found were predominantly M. varians, M. caseolyticus, and M. freudenreichii (in order of occurrence). Growth of micrococci with film yeasts indicated the yeasts stimulated the growth of the cocci on the surface of brick cheese smear. Butyric and acetic acids were identified as products of growth, but the fraction with the characteristic odor contained higher fatty acids.

Kaderavek et al. (1973) have shown that in milk products, lactic acid bacteria hydrolyze only short chain fatty acids. Micrococci from Italian cheeses are proteolytic and nonspecifically lipolytic, propionic acid bacteria lipases are nonspecific and product butyric, isovaleric, and valeric acids. They also found yeasts were lipolytic and mold lipases were nonspecific.

Summary

Flavors are generated from lipids by microbiological action. The microbiological action on lipids was shown for lactic acid bacteria, Micrococci, Staphylococci, Pseudomonas, yeast, and mold.

The lipase reactions for bacteria and mold are characteristically different for each type of organism and can differ by specificity within classes. Large differences in the lactic acid bacteria lipases were noted. Micrococci lipases are perhaps the most studied, and therefore the most defined.

Hydrolytic and oxidative reactions by the bacteria are, in general, ubiquitous regardless of product. The general scheme of an increase in free fatty acids, mono and di-carbonyls, and an increase in peroxide value, TBA, and total acids accompany the increase in free fatty acids.

The flavors of dry sausages and cheeses are dependent on the action of bacteria within and outside of the product. The flavors of cheeses and sausages are dependent upon (1) the specific organisms, (2) lipases produced by the organism, and (3) the substate furnished for the organism.

Literature Cited

1. Alford, J. A. and L. E. Elliott. Lipolytic activity of microorganisms at low and intermediate temperatures I. Action of Pseudomonas fluorescens on lard. Fd Res. (1960) 25, 296.
2. Alford, J. A. and D. A. Pierce. Lipolytic activity of microorganisms at low and intermediate temperatures III. Activity of microbial lipases at temperatures below 0°C. J. Food Sci. (1961) 26, 518.

3. Alford, J. A. and D. A. Pierce. Production of lipase by Pseudomones fragi in a synthetic medium. J. Bacti. 86, 24.
4. Alford, J. A., D. A. Pierce, and F. G. Suggs. Activity of microbial lipases on natural fats and synthetic triglycerides. J. Lipid Res. (1964) 5, 390-394.
5. Alford, J. A., D. A. Pierce, and W. L. Sulzbacker. Microbial lipases and their importance to the meat industry. Proc. Res. Conf. Advisory Council Res. Am. Meat Ind. Found. Univ. Chicago. (1963) 15, 11-16.
6. Alford, J. A. and J. L. Smith. Production of microbial lipases for the study of triglyceride structure. J. Am. Oil Chem. Soc. (1965) 42, 1038.
7. Alford, J. A., J. L. Smith, and H. D. Lilly. Relation of microbial activity to changes in lipids of foods. J. Appl. Bacteriol. (1971) 34, 133-146.
8. Aurand, L. W. and A. E. Woods. Food Chemistry, The AVI Publishing Company, Westport. (1973).
9. Blood, M. R. Lactic acid bacteria in marinated herring. "Lactic Acid Bacteria in Beverages and Food", Academic Press, New York, (1975).
10. Brockerhoff, H. Fatty acid distribution patterns of animal depot fats. Comp. Biochem. Physiol. (1966) 19, 1.
11. Cantoni, C., R. Molnar, P. Renon, and G. Giolitti. Lipids of dry sausages. Behavior of lipids during maturation. Ind. Conserve. (1966) 41, 188-197. CA. 66, 27809v (1967).
12. Cantoni, C., R. Molnar, P. Renon, and G. Giolitti. Micrococci in pork fat. J. Appl. Bacteriol. (1967) 30, 190-196.
13. Cerise, L., V. Bracco, I. Horman, T. Sozzi, and J. J. Wuhrmann. Changes in the lipid fraction during the ripening of pure pork salami; "Proceedings: 18th Annual Meeting of European Meat Research Workers", pp. 382-389. University of Guelph, Guelph, Ontario, Canada (1972).
14. Cerise, L., V. Bracco, I. Horman, T. Sozzi, and J. J. Wuhrmann. Changes in the lipid fraction during ripening of a pure pork salami. Fleischw 53, 223 (1973).
15. Collins, E. B. Biosynthesis of flavor compounds by microorganisms. "Journal of Dairy Science", 55, 1022-1028. (1972).
16. Cooke, B. C. Influence of incubation temperature on microbial lipase specificity. N.Z.J. Dairy Sci. Technol. 8, 126-127. (1973).
17. Coretti, K. The occurance and significance of lipolytic microorganisms in salami-type sausages intended for long storage. Fleischw. 45, 33. (1965).
18. Debevere, et al. Title unknown. Lebensmittel-Wissenschaft und Technologie. 9: 160. In the National Provisioner, 176, 83. (1976).

19. Deeth, H. C. and G. H. Fitz-Gerald. Lipolysis in dairy products, a review. "The Australian Journal of Dairy Technology", June, 53-64. (1976).
20. Demeyer, D., J. Hoozee, and H. Mesdom. Specificity of lipolysis during dry sausage ripening. J. Food Sci. 30, pp. 293-296. (1974).
21. Dobbertin, S., H. Siems, and H. Sinell. The bacteriology of fresh Mettwurst, II. Lipolytic activity and its dependency on the bacterial dynamics of fresh Mettwurst. Fleischw. 55, p. 237. (1975).
22. Fryer, T. F., R. C. Lawrence, and B. Reiter. Methods for isolation and enumeration of lipolytic organisms. J. Dairy Sci., 50, p. 447. (1967).
23. Fukuba, H. Studies on lipoxidase. Distribution of lipoxidase in plants and microorganisms. J. Agic. Chem. Soc. Japan. 26, p. 167. (1952).
24. Halvarson, H. Formation of lactic acid, volatile fatty acids, and neutral, volatile monocarbonyl compounds in Swedish fermented sausage. J. Food Sci. 38, 310-312. (1973).
25. Hawke, J. C. The formation and metabolism of methyl ketones and related compounds. J. Dairy Res. 33, p. 225. (1966).
26. Hornstein, I. and P. F. Crowe. Flavor studies on beef and pork. J. Agr. Food Chem. 8, pp. 494-498. (1960).
27. Hornstein, I. and P. F. Crowe. Meat flavor: lamb. J. Agr. Food Chem. II., pp. 147-149. (1963).
28. Jensen, R. G., J. Sampugna, J. G. Quinn, D. L. Carpenter, T. A. Marks, and J. A. Alford. Specificity of a lipase from Geotrichum candidum for cis-octadecenoic acid. J. Am Oil Chem. Soc., 42, p. 1029. (1965).
29. Kaderavek, G., S. Carini, and I. Saceinto. Effect of microbial lipases on milk fat. Riv. Ital. Sostanze Grasse. 50, pp. 135-136. CA. 79, p. 64582e. (1973).
30. Langer, H. J., H. Keckel, and E. Malek. Aroma substances in ripening dry sausage. Neutral monocarbonyl compounds in ripening dry sausages. Fleischw. 50, p. 193. (1970).
31. Lawrence, R. C., T. F. Fryer, and B. Reiter. The production and characterization of lipases from a micrococcus and a pseudomonad. J. Gen. Microbiol., 48, p. 401. (1967).
32. Lawrence, R. C. and J. C. Hawkes. The oxidation of fatty acids by mycelium of Penicillium roqueforti. J. Gen. Microbiol. 51, pp. 289-302. (1968).
33. Lu, J., and W. E. Townsend. Feasibility of adding freeze-dried meat in the preparation of fermented dry sausage. J. Food Sci., 38, pp. 837-840. (1973).
34. Lubert, D. J. and W. C. Frazier. Microbiology of the surface ripening of brick cheese. J. Dairy Sci., 38, pp. 981-990. (1955).

35. Mencher, J. R. and J. A. Alford. Purification and characterization of the lipase of Pseudomonas fragi. J. Gen. Microbial., 48, p. 317. (1967).
36. Mihalyi, V. and L. Kormandy. Changes in protein solubility and associated properties during the ripening of Hungarian dry sausages. Food Technol., 21, pp. 1398-1402. (1967).
37. Morris, H. A. and J. J. Jezeski. The action of micro-organisms on fats. II. Some characteristics of the lipase system of Pencillium roqueforti. J. Dairy Sci., 36, pp. 1285-1298. (1953).
38. Mukherjee, S. Studies on degradation of fata by microorganisms. I. Preliminary investigations on enzyme systems involved in the spoilage of fats. Arch. Biochem. 33, p. 364. (1951).
39. Nashif, S. A. and F. E. Nelson. The lipase of Pseudomonas fragi. II. Factors affecting lipase production. J. Dairy Sci., 36, p. 471. (1953).
40. Nurmi, E. Effect of bacterial inoculations on characteristics and microbial flora of dry sausages. Acta Agralia Fennica, p. 108. (1966).
41. Peterson, M. H. and M. J. Johnson. Delayed hydrolysis of butterfat by certain lactobacilli and micrococci isolated from cheese. J. Bact., 58, pp. 701-708. (1949).
42. Schelhorn, M. Cause of lipolysis in dry sausages studied with models. Fleischwirtschaft, 52, pp. 72-75. (1972).
43. Shimahara, K. Bacterial peroxidation of fats. III. Identification of lipoxygenase-forming bacteria. J. Ferment. Technol., Osaka. 44, p. 230. (1966).
44. Smith, J. L. and J. A. Alford. Action of microorganisms on the peroxides and carbonyls of rancid fat. J. Food Sci. 33, pp. 93-97. (1968).
45. Smith, J. L. and J. A. Alford. Action of microorganisms on the peroxides and carbonyls of fresh lard. J. Food Sci. 34, pp. 75-78. (1969).
46. Smith, J. L. and J. A. Alford. Inhibition of microbial lipases by fatty acids. Appl Microbial. p. 14, 699. (1966).
47. Stoychev, M. G. Djejava, and R. Brankova. Impact of pH and different NaCl concentrations on the lipase activity of starter cultures used in connection with the production of raw dried meat products. "Proceedings: 18th Annual Meeting of European Meat Research Workers", pp. 113-118. University of Guelph, Guelph, Ontario, Canada. (1972b).
48. Stoychev, M., G. Djejeva, and R. Brankova. Studies on lipase activity of some starter cultures at temperatures used in the manufacture and storage of raw-dried meat production. "Proceedings: 18th Annual Meeting of European Meat Research Workers", pp. 105-109. University of Guelph, Guelph, Ontario, Canada. (1972a).

49. Tappel, A. L. Lipoxidase. "The Enzymes", Vol. 8. Editors: P. D. Boyer, H. Lardy, and K. Myrback. Academic Press, New York. (1963).
50. Tjaberg, T. B., M. Haugam, and E. Nurmi. Studies on discoloration of Norwegian salami sausage. "Proceedings: 15th European Meeting of Meat Research Workers", August 17-24, 1969, Helsinki, Finland, pp. 138-148. (1969).
51. Wood, J. B., O. S. Cardenas, F. M. Yong, and D. W. McNulty. Lactobacilli in production of soy sauce, sour-dough bread, and Parisian barm. "Lactic Acid Bacteria in Beverages and Food", Academic Press, New York. (1975).

RECEIVED December 22, 1977

INDEX

G

H

I

V

W

Y